EMBRYOLOGY EXPLAINED

Embryology Explained is an essential guide for medical students and residents, enriched with original illustrations by Dr Thomas Newman that navigate the complexities of embryonic development. Structured methodically, the textbook breaks down embryogenesis with clarity and precision. Each chapter is organised into four main sections – embryology, clinical relevance, molecular details, and key points – allowing learners to efficiently review and connect the material with their specific academic and examination requirements. Tailored specifically to address the components of embryology emphasised in both undergraduate and postgraduate medical examinations, this book serves as an invaluable tool for examination preparation. It distinguishes itself from existing texts by shedding the common overly detailed approach and instead highlights the practical clinical applications of embryology, making it more pertinent and accessible for medical students and residents. It is an easy-to-read and comprehensive textbook that will explain embryology for you!

EMBRYOLOGY EXPLAINED

A COMPREHENSIVE TEXTBOOK FOR MEDICAL STUDENTS AND & RESIDENTS

SECOND EDITION

Fawz Kazzazi MB BChir, LLM, MBA, MA (Cantab), MRCS, FRSA
Plastic and Reconstructive Surgery Registrar (NTN London)
PhD Candidate – Medical Data Law

Diana Kazzazi MB ChB, MSc, MRCS
Plastic and Reconstructive Surgery Registrar (NTN North-East)

Danny Kazzazi MBBS, BSc (Hons), PgDip (ClinEd)
Specialized Foundation Programme Doctor

With Illustrations by
Thomas Hedley Newman
Urological Surgery Registrar (NTN South London)
Department of Urology, King's College Hospital, London, UK

CRC Press
Taylor & Francis Group
Boca Raton London New York

CRC Press is an imprint of the
Taylor & Francis Group, an **informa** business

Second edition published 2025
by CRC Press
2385 NW Executive Center Drive, Suite 320, Boca Raton, FL 33431

and by CRC Press
4 Park Square, Milton Park, Abingdon, Oxon, OX14 4RN

CRC Press is an imprint of Taylor & Francis Group, LLC

© 2025 Fawz Kazzazi, Diana Kazzazi, and Danny Kazzazi

This book contains information obtained from authentic and highly regarded sources. While all reasonable efforts have been made to publish reliable data and information, neither the author[s] nor the publisher can accept any legal responsibility or liability for any errors or omissions that may be made. The publishers wish to make clear that any views or opinions expressed in this book by individual editors, authors or contributors are personal to them and do not necessarily reflect the views/opinions of the publishers. The information or guidance contained in this book is intended for use by medical, scientific or health-care professionals and is provided strictly as a supplement to the medical or other professional's own judgement, their knowledge of the patient's medical history, relevant manufacturer's instructions and the appropriate best practice guidelines. Because of the rapid advances in medical science, any information or advice on dosages, procedures or diagnoses should be independently verified. The reader is strongly urged to consult the relevant national drug formulary and the drug companies' and device or material manufacturers' printed instructions, and their websites, before administering or utilizing any of the drugs, devices or materials mentioned in this book. This book does not indicate whether a particular treatment is appropriate or suitable for a particular individual. Ultimately it is the sole responsibility of the medical professional to make his or her own professional judgements, so as to advise and treat patients appropriately. The authors and publishers have also attempted to trace the copyright holders of all material reproduced in this publication and apologize to copyright holders if permission to publish in this form has not been obtained. If any copyright material has not been acknowledged please write and let us know so we may rectify in any future reprint.

Except as permitted under U.S. Copyright Law, no part of this book may be reprinted, reproduced, transmitted, or utilized in any form by any electronic, mechanical, or other means, now known or hereafter invented, including photocopying, microfilming, and recording, or in any information storage or retrieval system, without written permission from the publishers.

For permission to photocopy or use material electronically from this work, access www.copyright.com or contact the Copyright Clearance Center, Inc. (CCC), 222 Rosewood Drive, Danvers, MA 01923, 978-750-8400. For works that are not available on CCC please contact mpkbookspermissions@tandf.co.uk

Trademark notice: Product or corporate names may be trademarks or registered trademarks and are used only for identification and explanation without intent to infringe.

ISBN: 9781032766645 (hbk)
ISBN: 9781032766669 (pbk)
ISBN: 9781003479529 (ebk)

DOI: 10.1201/9781003479529

Typeset in Minion Pro
by Evolution Design & Digital Ltd.

DEDICATION

To our parents,

This book is dedicated to you: our first teachers, our steadfast supporters, and our inspirations. In every chapter, we see reflections of the sacrifices you made, the values you instilled, and the endless love you poured into nurturing us.

CONTENTS

	Preface to the Second Edition	viii
	Authors	ix
1	Basic Concepts	1
2	Sperm and Egg Formation	8
3	Fertilisation and Implantation	18
4	Gastrulation and Formation of the Axes	29
5	Segmentation and Folding	38
6	Neurulation and Brain Development	46
7	Craniofacial Development	57
8	Ear and Nose Development	70
9	Heart and Vessel Development	78
10	Limb Development	94
11	Bone Development	100
12	Foregut: Lung and Diaphragm	108
13	Foregut: Oesophagus and Stomach	117
14	Foregut: Hepatobiliary and Pancreas	123
15	Midgut Development	129
16	Hindgut and Bladder Development	137
17	Mesoderm: Spleen and Urinary System	144
18	Mesoderm: Internal and External Genitalia	154

PREFACE TO THE SECOND EDITION

Embryology is a challenging subject. A comprehensive knowledge of the processes that govern embryo formation can help medical students and doctors appreciate the mechanisms behind the pathologies they come across in clinical practice. Mastery of this subject not only deepens our understanding of human development but also provides critical insights into the aetiology of numerous pathologies across multiple specialties.

The difficulty with embryology teaching is that a good understanding is required to fully grasp the complex mechanisms. However, this depth of knowledge has not been matched in teaching schedules in medical schools worldwide. This, coupled with low representation in exams, has diminished the priority of the subject to the modern doctor.

This textbook is designed to simplify and demystify the intricate concepts of embryology, making them accessible and engaging. Our goal is for readers to not only grasp the fundamental mechanisms behind embryonic development but also to recognise the beauty inherent in the formation of human life.

Good luck and enjoy!

AUTHORS

Fawz Kazzazi is a Plastic and Reconstructive Surgery Registrar working in London. He has been teaching embryology for ten years at University of Cambridge Colleges having intercalated in the subject while attending medical school there.

Diana Kazzazi is a Plastic and Reconstructive Surgery Registrar working in Newcastle upon Tyne. She has been involved in medical education through her medical career and has developed workshops for residents, as well as completing a postgraduate certificate in medical education.

Danny Kazzazi is an Academic Foundation Doctor in a surgery-themed programme in London. He has been responsible for running nationwide and local teaching programmes as well as lectures.

1 BASIC CONCEPTS

Embryology encompasses the processes and stages involved in changing an undifferentiated cell into a multi-organ structure. It involves key cell interactions that control the differentiation and architecture of organ systems. This chapter will explore the core concepts that are essential to understanding embryology.

PLANES AND AXES

Position within the embryo is described using a series of terms that specify a structure's position: (1) relative to other structures, (2) relative to key landmarks, and (3) along three-dimensional axes. These terms can be confusing as some are relative while others are fixed, and they can vary according to species (Figure 1.1). They are listed below:

- *Rostral:* proximity to the head
- *Caudal:* proximity to the 'tail' or feet
- *Dorsal:* lies to the back or spine
- *Ventral:* lies to the front
- *Medial:* lies closer to the midline of the trunk or limb
- *Lateral:* lies further from the midline of the trunk or limb
- *Proximal:* lies closer to the trunk
- *Distal:* lies further from the trunk
- *Inferior:* below; for the bipedal human this will be similar to caudal
- *Superior:* above; for the bipedal human this will be similar to cranial
- *Anterior:* to the front – for the bipedal human this will be the same as ventral; for a four-legged/marine animal this will be equivalent to rostral
- *Posterior:* to the back – for the bipedal human, this will be the same as dorsal; for a four-legged/marine animal this will be equivalent to caudal

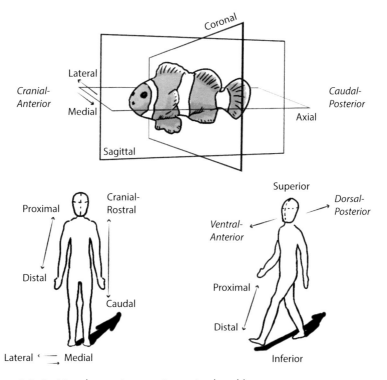

Figure 1.1 Positional terms in a marine animal and human.

CELLULAR PROCESSES

In order for undifferentiated cells to organise and become complex structures, they undergo five key cellular processes:

- *Differentiation:* the process during which embryonic cells specialise and diverse tissue structures arise; the form and function of the cell will typically change
- *Division:* the separation of a cell, often into equal identical parts
- *Multiplication:* the division of a cell into further cells
- *Chemotaxis:* chemical attraction of cells, leading to movement or differentiation of cells
- *Convergent extension:* shortening in one axis and lengthening in another (Figure 1.2)

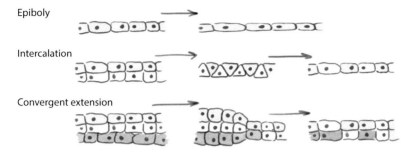

Figure 1.2 Depiction of the forms of axis extension in embryos. Epiboly involves individual cell changes, intercalation is the interposition of cells, and convergent extension is a form of cell rearrangement.

In order to achieve these cellular processes in the embryo, cells utilise *morphogens* and *actin–myosin contraction*. A morphogen is a molecule whose non-uniform distribution (typically along a gradient) is detected by cells. The concentration is interpreted by cells to establish their position relative to other cells and give instructions for differentiation. Actin–myosin contraction occurs in individual cells to allow them to move towards, or in response to, a morphogen.

UNDERSTANDING TIMELINES

The development of the embryo occurs in a near-uniform order, and this is sometimes confused with meaning that it always occurs at a specific time (e.g. day 10). For some, it is useful to consider the sequential events of fetal development through the 23 *Carnegie stages* that describe embryological features and events, rather than dates. The number of weeks of embryonic development differs from the *gestational age* (*GA*) that is described in clinical practice. The GA is calculated from the first day of the last menstrual period (typically adding 2 weeks to the embryonic development time).

The first 8 weeks of development (up to GA 10) are the *embryonic* period of development. Weeks 9 to 37 (GA 11 to 39) are the *fetal* periods of development. The first 4 weeks after birth are the *neonatal* period. Broadly, it can be said that the majority of organogenesis occurs in the first trimester, the second trimester involves developing and fine-tuning these organs, while the third trimester is the phase of rapid growth and preparation for life outside the uterus. As such, embryogenesis is considered to be most vulnerable during the first trimester, where the processes of organogenesis can be affected by medication, alcohol, recreational drugs, infectious agents, tobacco, radiation, and other chemical agents.

CLINICAL SIGNIFICANCE

An interruption to or alterations of the steps of organogenesis will lead to pathologies in the newborn, some of which are compatible with life. For each chapter, this book will outline the pathologies relevant to the processes described.

What you may notice is that the pathologies often exist along a spectrum from mild defects (which may not influence function) to complete atrophy/absence of an organ. Some defects are symptomless while others require immediate or delayed intervention. The 'Clinical Significance' section includes information to help identify both the symptoms (that the patient will report) and the clinical signs (that you will elicit on examination) from your future patients.

Fetal Warfarin Syndrome

Exposure to warfarin during pregnancy, especially in the first trimester, can significantly impact fetal development in a dose-dependent manner. This anticoagulant drug is known for its teratogenic effects, posing a heightened risk of inducing genetic mutations and a broad array of developmental abnormalities. The spectrum of warfarin-induced anomalies predominantly affects facial and skeletal structures, leading to conditions such as scoliosis, nasal hypoplasia, brachydactyly (shortening of the fingers), a shortened neck, and various chest deformities. Moreover, it is associated with a range of congenital heart defects and central nervous system disorders, including hydrocephaly, seizures, and microcephaly.

Warfarin's ability to cross the placental barrier and interfere with vitamin K metabolism underlies these adverse outcomes. Vitamin K plays a critical role in blood coagulation and bone growth, contributing to the synthesis of clotting factors II, VII, IX, and X. Additionally, it is essential for the proper formation and function of osteocalcin, a molecule secreted by osteoblasts that is pivotal for bone mineralisation and maturation. The disruption of vitamin K activity by warfarin, therefore, compromises both clotting mechanisms and skeletal development, underscoring the need for cautious management of anticoagulant therapy during pregnancy. Patients who are pregnant and require warfarin undergo specialist review and may be prescribed an anticoagulant that does not cross the placental barrier, such as heparin.

Fetal Alcohol Syndrome

Fetal alcohol syndrome (FAS) represents the most severe condition within a spectrum of disorders arising from maternal alcohol consumption during pregnancy. For a definitive diagnosis of FAS, the presence of the following criteria is essential:

- *Growth deficiency:* The fetus must exhibit growth measurements (either weight or height) below the tenth percentile, indicating significant prenatal growth restriction.
- *Central nervous system damage:* There must be evidence of structural or functional impairment within the central nervous system, which can manifest in a variety of developmental and cognitive disabilities.
- *Confirmed alcohol exposure in utero:* A documented history of maternal alcohol use during pregnancy is required, establishing a direct link between alcohol consumption and the observed symptoms.
- *Characteristic facial features:* The diagnosis also necessitates the identification of specific facial anomalies that are dose-dependent and include:
  - A smooth philtrum, which is the vertical groove between the base of the nose and the border of the upper lip
  - A thin upper lip, particularly noted in the vermilion, the red part of the lips
  - Small palpebral fissures, indicating shorter-than-normal openings for the eyes

Neonatal Abstinence Syndrome

This occurs when babies begin to withdraw from drugs that they are exposed to *in utero*. This can be seen in the offspring of mothers who take opiates during pregnancy. The baby will begin displaying typical syndromes of withdrawal, such as excessive crying, tremors, sleep disturbances, feeding difficulties, and heightened irritability, within the first few days after birth.

RELEVANT MOLECULES

For each chapter, there will be a summary of all the described cell signals, proteins, genes, and morphogens.

KEY POINTS

The 'Key Points' section will list a summary of the high-yield facts from the chapter that regularly appear in written examinations.

- Embryology is the processes and stages involved in changing an undifferentiated cell into a multi-organ structure.
- There is a subtle difference between cellular division and multiplication. Division involves the process of equal separation, whereas multiplication is the process of identical duplication.
- Organogenesis predominantly occurs in the first trimester.
- The embryonic period is the first 8 weeks of development. This is the same as the tenth week of the GA, as the GA incorporates the 2 weeks since the last menstrual period.
- A symptom is an experience that a patient describes.
- A sign is a clinical finding that can be elicited on examination.

2 SPERM AND EGG FORMATION

Gametogenesis is a fundamental biological process by which gametes, or sex cells, are produced in sexually reproducing organisms. This intricate process ensures that each gamete carries just half the genetic material of the parent, a condition essential for the maintenance of species-specific chromosome numbers across generations. In humans, this involves the formation of male gametes known as spermatozoa (sperm) through spermatogenesis, and female gametes called oocytes (eggs) through oogenesis. The mechanisms of gametogenesis are pivotal not only for reproduction but also for understanding genetic inheritance, congenital anomalies, and certain aspects of infertility. This chapter delves into the complexities of spermatogenesis and oogenesis, highlighting their unique features, stages, and the significant role they play in human development.

SPERMATOGENESIS

Spermatogenesis is the process by which spermatozoa are produced from *spermatogonial stem cells* (SSCs) in the seminiferous tubules of the testes (Figure 2.1). This highly regulated process can be divided into three main phases: the *proliferative* or *spermatogonial phase*, the *meiotic phase*, and the *spermiogenic phase*. These processes occur within the seminiferous tubules, which form the main tissue within the testicles and are responsible for sperm production. They are lined with a complex stratified epithelium that contains two cell types: spermatogenic cells and Sertoli cells. The former will develop into the spermatozoa; the latter provide an important nourishment role in spermatogenesis, and are also known as 'mother' or 'nurse' cells.

In the first phase (proliferative/spermatogonial phase), SSCs located at the periphery of the seminiferous tubules undergo a process of self-renewal and differentiation to form two types of cells. They differentiate into Type A (undifferentiated) and Type B (differentiated) cells. Type A cells continue to serve as stem cells to replenish the reservoir of SSCs, while the Type B cells are spermatogonia, which divide into *primary spermatocytes* ready for meiosis.

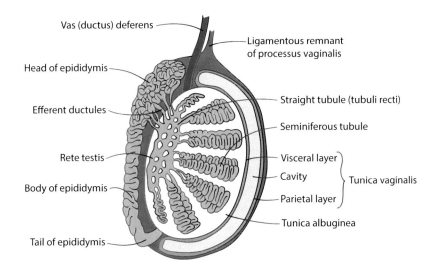

Figure 2.1 Structure of the testicle.

The Type B cells then proceed to the *meiosis* phase. These cells undergo the first meiotic division (Meiosis I) to convert diploid primary spermatocytes into haploid secondary spermatocytes. By way of reminder, Meiosis I is the first of the two consecutive divisions in meiosis, a specialised type of cell division unique to the production of gametes (sperm and eggs in animals). It is distinguished from mitosis and the second meiotic division (Meiosis II) by reducing the chromosome number by half, thus producing cells that are haploid (n), which means they contain one set of chromosomes. This reduction is crucial for sexual reproduction, ensuring that when gametes fuse during *fertilisation*, the resulting zygote has the correct diploid ($2n$) chromosome number, with one set of chromosomes from each parent. This process is preceded by interphase (DNA replication) and involves prophase (chromosome condensation and genetic recombination), metaphase I (chromosome alignment and randomisation), anaphase I (separation of chromosomes), telophase I (formation of new nuclei), and cytokinesis (formation of new cells).

Once the primary spermatocytes have undergone Meiosis I to form *secondary spermatocytes*, they then undergo Meiosis II to produce haploid *spermatids*. The steps in Meiosis II are very similar to those in Meiosis I. The overall reductional division is crucial as it halves the chromosome number from 46 (diploid) to 23 (haploid), ensuring that each sperm carries only one set of chromosomes. However, unlike Meiosis I, which reduces the chromosome number by half (reductional division), Meiosis II separates the sister chromatids of each chromosome in a manner similar to a mitotic division (equational division). This results in four

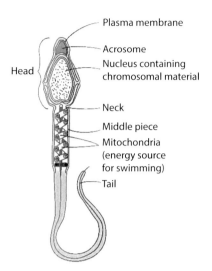

Figure 2.2 Structure of a mature spermatozoon.

haploid cells, each with a single set of chromosomes from the original diploid cell. Meiosis II is crucial for ensuring that each gamete contains exactly one complete set of chromosomes, a necessity for maintaining the species-specific chromosome number upon fertilisation.

Finally, in the spermiogenic phase, spermatids undergo a series of morphological and physiological changes to become mature *spermatozoa* (Figure 2.2). This includes the development of a flagellum for motility, condensation of nuclear material to facilitate DNA transport, and formation of the acrosome, which is essential for fertilisation. The acrosome is a cap-like structure which contains hydrolytic enzymes that allow the sperm to penetrate the glycoprotein membrane of the oocyte. Once the acrosome has formed, the sperm cell then undergoes a process of cytoplasm reduction and shedding in order to become more streamlined. The entire process of spermatogenesis takes about 64 days, resulting in sperm that are capable of fertilising an oocyte.

OOGENESIS

Oogenesis is the process by which oocytes are produced in the ovaries (Figure 2.3). Unlike spermatogenesis, oogenesis involves not only cell division but also a significant period of growth and maturation that spans a female's reproductive lifespan. Oogenesis can be divided into several phases: *oogonial, follicular, ovulatory*, and *post-ovulatory/fertilisation*.

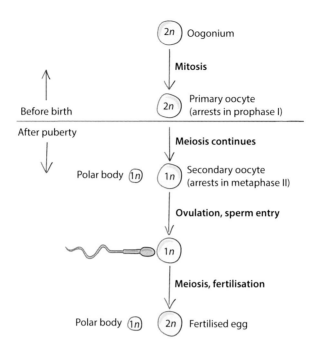

Figure 2.3 Summary diagram of oogenesis.

The oogonial phase begins in early embryogenesis in the fetal ovaries. *Oogonia* develop from the primordial germ cells (PGCs) and multiply to increase in number. By mid-gestation, the oogonia will become *primary oocytes* after entering the first meiotic division. Remarkably, this entry into Meiosis I is unique as it halts at the prophase I stage, where it remains arrested until puberty. This arrest can last for decades, a feature peculiar to human oogenesis.

This is then followed by a follicular phase, where the primary oocytes will mature under regulation by hormones. Concurrent with the arrest in prophase I, in a process known as *folliculogenesis*, primary oocytes are surrounded by a layer of somatic cells – which will become granulose cells – to form primordial follicles. This encapsulation marks the beginning of folliculogenesis. Thus, by the time of birth, a female's ovaries contain her lifetime supply of oocytes, each encased within a *primordial follicle*. These primordial follicles remain in a dormant state until puberty, when hormonal changes trigger the beginning of the ovarian cycle, leading to the maturation of some of these follicles (under the direction of *follicle-stimulating hormones* [FSH]).

At each menstrual cycle, a select group of primary oocytes resumes Meiosis I in the ovulatory phase, but only one typically reaches full maturity. This is triggered by a surge in luteinising hormone (LH). When the mature primary oocyte completes

Meiosis I, it produces a *secondary oocyte* and a smaller polar body, then enters Meiosis II and is arrested in metaphase II, awaiting fertilisation. If fertilisation occurs, the secondary oocyte completes Meiosis II, dividing into a haploid ovum and another polar body. The ovum then merges with a spermatozoon to form a *zygote*, marking the beginning of a new life.

In contrast to spermatogenesis, which occurs continuously and produces millions of sperm daily, oogenesis is cyclical and results in the release of a single mature oocyte per menstrual cycle. This process emphasises the precise regulatory mechanisms in place to ensure the production of viable gametes for reproduction. Together, spermatogenesis and oogenesis are essential for sexual reproduction, ensuring genetic diversity and the continuation of species. Understanding these processes provides insight into human development, reproductive health, and the basis for certain infertility treatments.

CLINICAL SIGNIFICANCE

The processes of oogenesis and spermatogenesis are complex and highly regulated, making them susceptible to various disruptions that can lead to clinical conditions affecting fertility and reproductive health. Understanding these conditions is crucial for the diagnosis, management, and treatment of reproductive disorders.

Oligospermia and Azoospermia

Oligospermia refers to a low sperm count, while azoospermia is the absence of sperm in the ejaculate. Both conditions can be caused by various factors, including hormonal imbalances, genetic defects, and exposure to toxins, leading to difficulties in achieving pregnancy. According to the World Health Organization, oligospermia is defined as fewer than 15 million sperm per millilitre of semen.

Azoospermia is a more severe condition characterised by the complete absence of sperm in the ejaculate. It affects about 1% of all men and 10–15% of men with infertility. It can be divided into obstructive and non-obstructive causes, the former being related to blockages in the ductal system and the latter due to poor sperm production.

Varicocele

A varicocele is an enlargement of the veins within the scrotum. It can decrease sperm production and quality, potentially leading to infertility. Varicoceles are associated with impaired temperature regulation in the testes, affecting spermatogenesis. When a large varicocele is detected during clinical examination, it is often described as feeling like a 'bag of worms'. Approximately 15% of all men have varicoceles and it is usually both asymptomatic and without clinical effect.

Klinefelter Syndrome

This genetic condition occurs when a male is born with an extra X chromosome (XXY instead of XY; known as 47XXY). Men with Klinefelter syndrome may have smaller testes, which can lead to reduced levels of *testosterone* and impaired spermatogenesis, often resulting in infertility. The smaller testes arise due to issues in the development of both the seminiferous tubules and the Leydig cells (the latter of which causes reduced testosterone production). This leads to a reduction in the ability to produce sperm, the number of seminiferous tubules, and the development of the testes. Ultimately, this can lead not only to smaller testes but also to fibrosis and hyalinisation, whereby the testes become scarred with connective tissue. This condition is associated with other clinical signs (often noticed at puberty with reduced testosterone production), including reduced libido, gynaecomastia, reduced body hair, and generalised weakness.

▌ Cryptorchidism

Also known as undescended testicles, this condition occurs when one or both of the testes fail to descend from the abdomen into the scrotum before birth. Normally, testicles descend into the scrotum during the last trimester of fetal development. Cryptorchidism is one of the most common congenital anomalies in males, affecting about 2–5% of full-term and up to 30% of premature male infants. However, in most cases, the testes descend spontaneously within the first few months of life.

A history of this condition accounts for up to 10% of all infertility in men. Cryptorchidism can lead to impaired spermatogenesis and increased risk of infertility due to the increased temperature of the testicle. The primary goal of treatment for cryptorchidism is to reduce the risk of complications, including infertility, testicular cancer, testicular torsion, and psychosocial concerns related to an empty scrotum. Treatment typically involves surgical correction to move the undescended testicle into the scrotum, a procedure known as orchiopexy. This surgery is recommended for most cases of cryptorchidism and is ideally performed on infants aged between 6 months and 18 months. The earlier the procedure is done, the better the chances for normal testicular development and fertility. During orchiopexy, the surgeon carefully mobilises the undescended testis, bringing it down into the scrotum, and secures it in place.

▌ Premature Ovarian Failure

Premature ovarian failure (POF), also known as primary ovarian insufficiency, is a condition characterised by the loss of normal ovarian function before the age of 40 years. Unlike menopause, which is a natural part of ageing, POF happens earlier and can lead to a decrease in *oestrogen* levels and fewer eggs within the ovaries. The exact causes of POF are varied and often difficult to pinpoint, but they can include a combination of genetic factors, autoimmune diseases, environmental exposures, and medical treatments.

▌ Polycystic Ovary Syndrome

Polycystic ovary syndrome (PCOS) is a complex endocrine disorder commonly affecting 5–10% of women of reproductive age. Women with PCOS may have infrequent or prolonged menstrual periods or excess male hormone (androgen) levels. The ovaries may develop numerous small collections of fluid (follicles) and fail to regularly release eggs, leading to ovulatory dysfunction and infertility.

It is marked by the presence of insulin resistance (due to hyperinsulinaemia), hormonal imbalances (elevated levels of androgens), and low-grade inflammation. Its clinical manifestation varies according to its severity with signs relating to these key features. For example, pertaining to the hyperandrogenism, patients can present with acne, hirsutism (excess hair), and androgenic alopecia (male pattern hair loss). The hormonal imbalance also has impacts on the menstrual cycle ranging from amenorrhoea (absence of periods), oligomenorrhoea (infrequent periods), to irregular periods.

A diagnosis of PCOS is made if there is satisfaction of at least two out of the following three criteria, known as the Rotterdam criteria:

- Menstrual period absence or irregularity
- Clinical or biochemical hyperandrogenism
- Polycystic ovaries on ultrasound

The management of PCOS involves targeting the key factors with hormonal contraceptives, anti-androgens (e.g. spironolactone), metformin, lifestyle changes leading to weight loss, and fertility treatments.

Turner Syndrome

Turner syndrome is a chromosomal disorder affecting females, characterised by the partial or complete absence of one of the two X chromosomes, resulting in a 45X0 karyotype or other variations involving the X chromosome. Occurring in approximately 1/2500 live female births, Turner syndrome leads to a range of developmental and medical issues. Key features include short stature, which often becomes evident by the age of 5 years, and ovarian insufficiency leading to POF and infertility in the majority of cases.

Additional common physical manifestations can include webbed neck, low hairline at the back of the neck, lymphoedema of the hands and feet, and skeletal abnormalities. Cardiovascular anomalies, such as bicuspid aortic valves and coarctation of the aorta, are significant concerns and require ongoing monitoring. Hormone replacement therapy is typically necessary to promote sexual development and maintain bone health.

RELEVANT MOLECULES

- *FSH:* a gonadotropin released by the anterior pituitary gland that stimulates follicle growth and maturation in ovaries and spermatogenesis in testes
- *LH:* another gonadotropin produced by the anterior pituitary gland, LH triggers ovulation and the development of the corpus luteum in females and stimulates testosterone production in males
- *Testosterone:* a steroid hormone produced primarily in the testes, which is crucial for the development of male secondary sexual characteristics and spermatogenesis
- *Oestrogen:* a group of steroid hormones, predominantly produced in the ovaries, that promote the development and maintenance of female characteristics and regulate the menstrual cycle
- *Progesterone:* a steroid hormone released by the corpus luteum and the placenta, which prepares the endometrium for pregnancy and maintains the uterine lining during early pregnancy
- *Gonadotropin-releasing hormone (GnRH):* secreted by the hypothalamus, GnRH controls the release of FSH and LH from the anterior pituitary gland

KEY POINTS

- Spermatogenesis is a continuous process beginning at puberty, occurring in the seminiferous tubules of the testes.
- Spermatogenesis involves three phases – spermatogonial (mitotic), meiotic, and spermiogenic – resulting in the formation of mature spermatozoa.
- Oogenesis begins prenatally with the formation of primordial follicles, each containing a primary oocyte arrested in prophase I of meiosis.
- Oogenesis resumes at puberty, with one selected follicle maturing each menstrual cycle, leading to ovulation of a secondary oocyte arrested in metaphase II.
- FSH stimulates the growth of ovarian follicles and spermatogenesis.
- LH triggers ovulation and testosterone production from Leydig cells.
- Testosterone is essential for spermatogenesis and the development of male secondary sexual characteristics.
- Oestrogen and progesterone regulate the menstrual cycle and prepare the endometrium for pregnancy.

3 FERTILISATION AND IMPLANTATION

This area of medicine is repeatedly assessed in both preclinical and clinical medicine as it contains important examinable knowledge for obstetrics, gynaecology, surgery, and emergency medicine.

STRUCTURES

Following ovulation, the female gamete (oocyte) is fertilised by the male gamete (sperm). The structures you need to be familiar with are:

- *Oocyte:* the female gamete (egg, ova, ovum) prior to fertilisation
- *Zona pellucida:* the protective extracellular glycoprotein matrix surrounding the oocyte
- *Zygote:* the fertilised (diploid) ovum
- *Morula:* following fertilisation and several rounds of mitotic division (to form identical cells), this structure is a ball of 16 cells (*morula* being Latin for 'mulberry')
- *Blastocyst:* the sphere of cells that forms after the morula, characterised by the existence of an inner fluid-filled cavity
- *Blastocoel:* the cavity that the cells of the blastocyst surround
- *Blastomeres:* the individual cells of the blastula
- *Blastopore:* an opening in the blastula through which gastrulation occurs (this process will be discussed in Chapter 4)
- *Trilaminar disc:* a structure that forms and marks that gastrulation has occurred, containing the three germ layers

TIMELINE

- *Day 1:* fertilisation
- *Days 1 to 3:* first cleavage (2–16 cells)
- *Days 4 to 6:* the blastocyst enters the uterus and implants
- *Days 7 to 12:* implantation is complete
- *Day 13:* formation of the primary stem villi and primitive streak
- *Day 16:* gastrulation

FERTILISATION

This occurs in the *ampulla* of the oviduct (Figure 3.1). Spermatic entry occurs by breaking through the oocyte's protective glycoprotein zona pellucida by binding with the ZP3 receptors. First, the sperm's membrane fuses with the outer layer of the oocyte, releasing *acrosomal enzyme* (contained within the acrosome in the head of the sperm). This contains degradative substances, such as *hyaluronidase* and *acrosin*, to digest the outer membranes and allow entry – a process known as the *acrosomal reaction*. The fusion of sperm and oocyte membranes triggers a series of calcium waves in the oocyte, which helps to complete the oocyte's second meiotic division. Following this, cleavage of the ZP3 receptors occurs on the zona pellucida to prevent the entry of further sperm (and additional chromosomes).

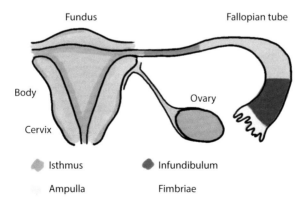

Figure 3.1 The labeled uterus and segments of the oviduct.

CLEAVAGE AND FORMATION OF THE BLASTOCYST

In mammals, the point of sperm entry determines the plane of the first division of the zygote. Key to the process of cleavage is that it occurs *without a change in size* – it is a pure division of cells into equal parts. The zygote will subdivide into blastomeres. At the eight-cell stage, the loose arrangement of blastomeres is compacted. When these blastomeres divide again into a 16-cell structure, it is called the morula (Figure 3.2). The morula will reach the uterus by around day 3 to 4.

Due to the size and structure of the morula, an inside-out axis (polarity) is formed. This is because there is now a difference between cells in the morula; they are not all the same. The innermost cells are surrounded by cells on all sides, whereas the outer cells only contact cells on their inner surface. Therefore, the cells do not receive equivalent signals as the innermost cells will receive signals from all sides, whereas the outer cells only receive cell signals on their inner surface. An *inside-out polarity* is formed. This polarity will assist in the differentiation of cells as they receive different signals to transform.

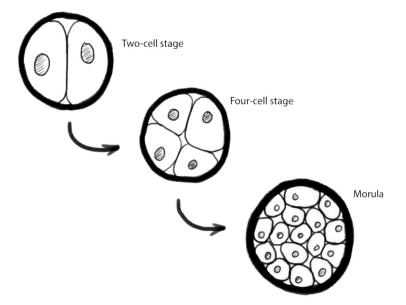

Figure 3.2 The two-cell zygote subdividing (without an increase in size) into the morula.

FORMATION OF THE INNER CELL MASS AND TROPHOBLAST

The inner cells of the morula will form the *inner cell mass* (ICM) and the outer cells will form the *trophoblast*. The ICM becomes the *embryo proper* (the cells which form the fetus). Due to the inside-out polarity, there is a discrepancy in the expression of sodium/potassium pumps. This leads to an influx of ions to the centre of the morula, through which fluid follows ('water follows salt'). This creates a cavity in the middle of the morula, and the structure is now the blastocyst (blastomeres surrounding a central blastocoel). This pushes the ICM to one end of the blastocyst (Figure 3.3) such that there are two poles: the embryonic pole (where the embryo proper is) and the anembryonic pole.

The ICM is *pluripotent* and can be cultured into embryonic stem cells. It is pluripotent rather that totipotent as it has partially differentiated to become the ICM and lost the ability to form the trophoblast (i.e. cannot form extra-embryonic cells). The ICM will differentiate into a bilaminar (two-layered) structure composed of the *epiblast* and the *hypoblast*. When these two layers form, the first asymmetry of the embryo forms, resulting in the *dorso-ventral axis*. This is because in becoming double-layered, the embryo now has two surfaces and thus has dorso-ventral asymmetry.

The trophoblast will become the *syncytiotrophoblast* (STB) and the *cytotrophoblast* (CTB). As the name suggests, the STB is a syncytium (similar to myocardial cells) because many cell nuclei are contained within a single cell membrane. The CTB (*cyto* meaning 'cell') is more like a typical cellular structure with a single nucleus per cell.

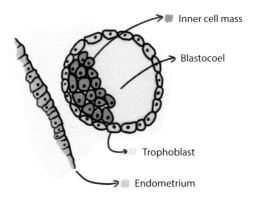

Figure 3.3 The blastocyst with an inner cell mass at the embryonic pole heading towards the endometrium of the uterus.

IMPLANTATION AND THE FORMATION OF THE AMNION

Implantation of the embryo is needed as oviductal nourishment (without a placenta, e.g. a chicken egg) is too poor to support human embryo development; therefore, a haemotrophic method of nutrient exchange is required. On day 5, the blastocyst will hatch from the zona pellucida; between days 6 and 7, the STB develops and begins to invade the endometrium. The STB does this, as opposed to the CTB, owing to its looser structure, which allows invasion of the maternal uterine wall.

Trophoblastic lacunae open up within the STB and nearby maternal capillaries expand to form *maternal sinusoids* that rapidly anastamose. Proliferation of local CTB provides more structure to the invading STB and vasculature to form the *primary chorionic stem villi*. This implantation process will lead to the expression of *beta human chorionic gonadotrophin* (β-HCG) by the trophoblast. These anastamoses will develop further to ultimately form the placenta; as the trophoblast proliferates, more β-HCG is produced.

At a similar time, an additional cavity will form between the ICM and the trophoblast at the embryonic pole: this is the *amnion* (Figure 3.4). Specifically, the amnion will form between the epiblast (which sits closer to the endometrium) and the CTB. Cells from the hypoblast will migrate around the blastocoel to form an

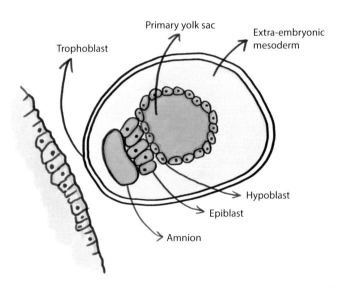

Figure 3.4 The embryo containing the epiblast and hypoblast (products of the ICM) and the amnion (between the epiblast and trophoblast).

inner layer; hence the hypoblast is sometimes called the 'primitive endoderm'. This will then form the *primary yolk sac* (extracoelomic cavity).

FORMATION OF THE YOLK SAC AND CHORIONIC CAVITY

Cells from this yolk sac will further divide within the CTB to form *extra-embryonic mesoderm* (EEM) ('extra-embryonic' as it will not form part of the embryo proper). The cells of the EEM will then separate into two forms: the *somatic (parietal) EEM* and *visceral EEM* (Figure 3.5). The visceral EEM is the inner layer (closer to the embryo proper) and the cavity that forms between the visceral and somatic EEM will be the *chorionic cavity* (chorion). After a further migration and lining with hypoblast cells, the primary yolk sac becomes the *secondary yolk sac*.

The EEM in humans forms the extra-embryonic membranes: the amnion, yolk sac, chorion, and allantois. The collective role of the yolk sac, chorion, and amnion is

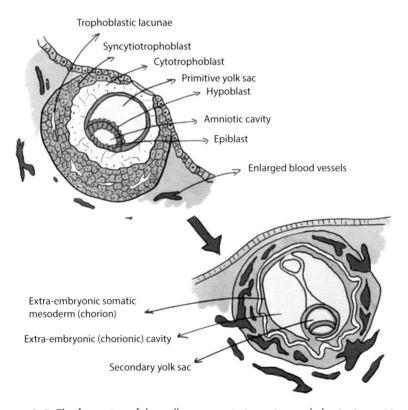

Figure 3.5 The formation of the yolk sac, amniotic cavity, and chorionic cavities.

nutrient uptake and excretion; these are of more importance in non-mammals that do not have placentas. The extra-embryonic membrane system involves a small diverticulum, known as the *allantois*, that functions to excrete waste materials from embryos. The amnion cells are secretory epithelial cells that contribute to the formation of *amniotic fluid*, which acts as a mechanical barrier to the embryo. The yolk sac is the first site of blood cell production and so contributes to early erythropoiesis, haematopoesis, and vasculogenesis. It also generates the *primordial germ cells* (which will be discussed later). The chorion will aid formation of the chorionic stem villi and contribute to the fetal portion of the placenta.

CLINICAL SIGNIFICANCE

▎ Ectopic Pregnancy

This occurs when the fertilised egg implants at a site other than the uterus. Most commonly, this happens in the fallopian tube. The likelihood of it occurring is increased in patients who have had previous operations, scarring, or chronic inflammation of their fallopian tubes. As the embryo cannot develop in this confined space, a rupture of the tubes can occur – this is an emergency.

The symptoms of ectopic pregnancy include abdominal pain (first general, then often localised to the site as it progresses), missed periods, bloody vaginal discharge, and discomfort on urination/defecation. If the fallopian tube ruptures then the abdominal pain becomes diffuse or peritonitic, and the patient may report shoulder-tip discomfort. This occurs because the bleeding from the site of rupture irritates the underside of the diaphragm, which is supplied by the phrenic nerve (roots C3, C4, C5), and the pain is referred to the corresponding dermatome.

▎ β-HCG Levels in Pregnancy

Pregnancy can be confirmed by detecting levels of β-HCG in the blood serum and the urine. This marker is produced by the trophoblast cells (the STB initially) after implantation. Furthermore, by monitoring the trend of the level in the serum, the viability of the pregnancy can be determined by comparing it to expected levels. The β-HCG levels are proportional to the volume of trophoblast/placental tissue. During the first 4 weeks of a viable pregnancy, the serum β-HCG level would be expected to double every 2–3 days as the trophoblast cells proliferate and the placenta grows. After 6 weeks, the level doubles every 4 days. In ectopic pregnancies, one can imagine that the trophoblasts and placenta cannot increase in size (and release β-HCG) due to a lack of viable tissue and space in the ectopic site. As such, if there is no doubling of the β-HCG serum level then the pregnancy may be ectopic or non-viable.

Such is the importance of diagnosing an ectopic pregnancy that every patient of childbearing potential and age presenting to the emergency department of a hospital with abdominal pain should have a urine β-HCG test to exclude the existence of an ectopic pregnancy, as a delay in operating can be fatal.

▎ Radiation Risk in Pregnant Patients

Healthcare professionals should consider the risk of ionising radiation to the fetus in pregnant patients. This can increase the risk of mutation during organogenesis and subsequent disease in the offspring. As such, the type of scan should be considered in all patients who could be pregnant to minimise these risks. This may take the form of selecting an ultrasound scan as opposed to a computed tomography (CT) scan (e.g. an ultrasound abdomen rather than CT abdomen–pelvis to investigate acute abdominal pain). As such, all patients of childbearing potential should have a urine pregnancy test prior to a scan.

RELEVANT MOLECULES

- *ZP3 receptors:* these are present on the zona pellucida; the sperm bind to these receptors to enter the oocyte
- *β-HCG:* hormone released by the trophoblast on implantation

KEY POINTS

- Fertilisation occurs in the ampulla of the oviduct.
- Division from zygote to blastomeres is mitotic.
- Division to the morula stage occurs *without* an increase in cell mass – there is subdivision of the zygote without a change in size.
- The ICM is pluripotent.
- The ICM will form the epiblast and hypoblast.
- The hypoblast is also known as the primitive endoderm.
- β-HCG is released by the trophoblast (mainly STB) and used as a serum and urine marker for pregnancy.
- Any patient of childbearing age and potential who presents to the emergency department of a hospital with abdominal pain should have a β-HCG urine test to rule out ectopic pregnancy.
- The yolk sac is the source of the primordial germ cells.
- The chorion contributes to the fetal aspect of the placenta.

4 GASTRULATION AND FORMATION OF THE AXES

Gastrulation is considered by many to be the most important step in embryogenesis. By the end of this process, the definitive body axes have been established and the basic architecture of human tissues has been created. It results in the formation of three germ layers (endoderm, mesoderm, and ectoderm) in a trilaminar disc, which will then undergo folding to generate the positioning of organs as we know them in the adult body.

GASTRULATION

Around the third week of development, the epiblast cells that lie *caudally* will condense to form the *primitive streak*. This marks the start of gastrulation. At the cranial end of this streak lies the *primitive pit* and the *primitive node*. The cells of the epiblast are *epithelial* and connected by a homodimer called *E-cadherin* (a strong adhesive molecule between epithelial cells). During gastrulation, these cells ingress at the primitive streak to create the three germ layers: *endoderm*, *mesoderm*, and *ectoderm*. In order to do this, the cells at the primitive streak release *fibroblast growth factor 8* (*Fgf8*) to suppress the E-cadherin adhesion between epiblast cells, so that they can mobilise in a process known as *epithelial-to-mesenchymal transition*.

Now, with more fluidity in the structure due to suppression of the connective bonds, these epiblast cells move towards the primitive streak then ingress through it to displace the underlying hypoblast (Figure 4.1). The first migrating epiblast cells will become the *definitive endoderm* (replacing the primitive endoderm [the hypoblast]). The next cells to ingress will become the mesoderm. The remaining cells of the epiblast form the ectoderm. This creates a structure composed of three layers – the trilaminar disc. A useful list of the derivatives of each germ layer is given in Figure 4.2.

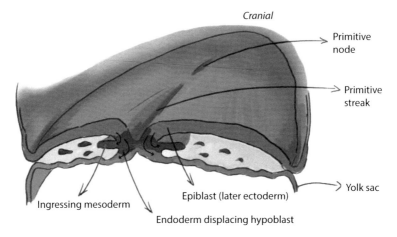

Figure 4.1 Epiblast cells ingressing at the primitive streak to displace the hypoblast and form the three germ layers.

ANTERO-POSTERIOR AXIS

At the cranial end of the embryonic disc, a portion of endoderm (the *anterior visceral endoderm*) modifies gene expression and morphogen levels to define the antero-posterior axis. It upregulates various transcription factors (*LIM1, HESX1, OTX1*) to secrete the morphogens *Lefty* and *Cerberus*. These inhibit the *Nodal* molecules secreted by the primitive streak. This creates a gradient from the cranial (anterior) end of the disc to the caudal (posterior) end. This ensures that the appropriate genes for structures of the head are expressed anteriorly. This can be proven by explanting the anterior visceral endoderm to another part of the embryo, and demonstrating the changes in gradient and subsequent structural development.

MESODERM SUBDIVISION

The mesoderm will divide into multiple subgroups of cells defined by their position and laterality to the notochord: the *paraxial, intermediate*, and *lateral plate mesoderm*. These are determined by the expression of signals from the notochord and the primitive streak.

Bone morphogenetic protein 4 (*Bmp4*) is secreted throughout the embryonic disc. It is antagonised in the midline by the primitive node's expression of *Chordin* and *Noggin*. This signals to the nearby mesoderm to form the dorsal structures of the *notochord* and paraxial mesoderm. The non-antagonised expression of Bmp4 will

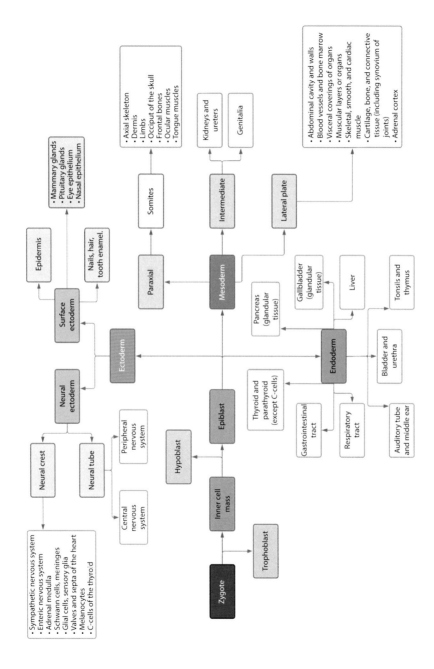

Figure 4.2 Quick reference guide to the germ layer origin of organs and tissues.

Gastrulation and formation of the axes 31

work with Fgf8 to signal to the remaining mesoderm to become the lateral plate and intermediate mesoderm.

DORSAL-VENTRAL AXIS

The formation of the three layers signifies the creation of the dorsal-ventral axis of the embryo. On a molecular level, this is due to Nodal in the primitive streak (positioned dorsally) upregulating a series of morphogens that stimulate cell differentiation and create different layers.

LEFT-RIGHT AXIS

The formation of the left-right axis is dependent on the primitive node, ciliated cells, and serotonin (5-hydroxytryptamine [5-HT]) expression. Ciliated cells waft morphogens onto the left side, signalling to Fgf8 on the left to upregulate expression of Nodal and *Lefty2* on this side and promote expression of 5-HT. This leads to left-sided features. On the right, *Snail* (a transcription factor) is upregulated and promotes right-sided genes; furthermore, uninhibited monoamine oxidase enzymes break down 5-HT to inhibit left-sided features.

CLINICAL SIGNIFICANCE

▎ Layers and Derivatives

The germ layer origin of organs or tissues is a topic frequently assessed in multiple-choice questions and spotter stations. Figure 4.2 acts as a quick reference guide. A simplification to assist educated guessing in examinations is that: the endoderm creates the innermost tube (e.g. the gut) and the organs that connect directly to it (e.g. the gallbladder); the ectoderm forms the outermost tissues (e.g. the epidermis of skin); and the mesoderm forms the tissues that lie between these two layers (e.g. the kidneys).

▎ Endoderm

Predominantly, this forms the gastrointestinal tube and structures that bud from or attach to it, including:

- Gastrointestinal tract (epithelial lining)
- Respiratory tract (epithelial lining)
- Auditory tube and middle ear
- Bladder and urethra
- Tonsils and thymus
- Liver glandular tissue
- Gallbladder
- Pancreatic glandular tissue
- Thyroid and parathyroid gland parenchyma (except C cells)

▎ Mesoderm

This generally forms organs that lie between the gastrointestinal tract and the epidermis:

- *Paraxial*
  - Somites
    - Axial skeleton (vertebrae and ribs) with overlying dermis and associated muscles (NOT the epidermis)
    - Limbs
    - Occipital part of the skull
    - Frontal bones (eyes, nose, inner ear)
    - Ocular muscles
    - Tongue muscles

- *Intermediate mesoderm*
  - Kidneys
  - Ureters
  - External (gonads) genitalia and internal (reproductive) genitalia
- *Lateral plate mesoderm*
  - Abdominal cavity and walls
  - Blood vessels and bone marrow
  - Visceral coverings of organs
  - Muscular layers of organs
  - Skeletal, smooth, and cardiac muscle
  - Cartilage, bone, and connective tissue (including synovium of joints)
  - Adrenal cortex

Ectoderm

Broadly, the ectoderm forms sensory and epidermal (most external) structures, including:

- Peripheral nervous system
- Cranial nervous system
- Epidermal skin including receptors and hair follicles
- Nails
- Teeth (Note: odontoblasts have neural crest origin, whereas the enamel does not; teeth developed in vertebrates to allow fish to detect acidity/temperature of water)
- Mammary glands
- Pituitary gland
- Eyes
- Sensory epithelium of nose
- Neural crest cells
  - Sympathetic nervous system
  - Enteric nervous system
  - Adrenal medulla
  - Schwann cells
  - Meninges
  - Glial cells
  - Valves/septum of the heart

- Melanocytes
- Sensory ganglia
- C cells of thyroid

▍Situs Inversus Totalis

This condition occurs in about 1/10,000 people and is characterised by the complete mirroring of the organs such that left-sided structures are on the right (and vice versa); for example, the liver is positioned on the left instead of the right. It occurs most commonly in an autosomal recessive pattern and patients are typically asymptomatic. The specific mutated genes are heterogenous and vary between families. In 10% of cases there are concurrent congenital heart conditions which can be severe.

Clinically, this condition may be encountered when patients present with atypical unilateral symptoms; for example, migratory left-sided abdominal pain consistent with appendicitis.

▍Dextrocardia and Levocardia

In patients with dextrocardia, the heart is found on the right side of the thorax; it can occur as part of situs inversus or in isolation. For practical management of these patients, electrocardiogram leads should be mirrored. When occurring in isolation, it can be associated with defects of the heart and/or lungs, such as valve or septal malformations.

Levocardia means the apex of the heart is left-sided – which is normal. It is a useful term in cases where other organs are mirrored to explain the position of the heart; for example, situs inversus with levocardia.

▍Primary Ciliary Dyskinesia and Kartagener's Syndrome

As discussed previously, ciliated cells waft molecules for early determination of left-right axis. Where there is a dysfunction in the motility of motor proteins in ciliated cells in patients, they are said to have primary ciliary dyskinesia (PCD). As such, patients with PCD have a 50% chance of developing situs inversus as the determination molecules have an equal chance of lying either side of the midline, owing to not being wafted by ciliated cells. PCD is an autosomal recessive disease and affects organs where flow or function is dependent on the efficient motility of cilia. As such, patients have defects in their respiratory tract, middle/inner ears, sinuses, and reproductive tracts/cells (e.g. sperm ejaculation). This manifests in conditions such as bronchiectasis, infertility, chest and ear infections, hearing loss, and sinusitis.

In Kartagener's syndrome, patients have a triad of situs inversus, chronic sinusitis, and bronchiectasis.

RELEVANT MOLECULES

- *E-cadherin:* strong adhesive molecule between epithelial cells that is suppressed by Fgf8 in the epiblast cells to allow gastrulation
- *LIM1, HESX1, OTX1, Cerberus, Lefty:* upregulated by the anterior visceral endoderm to define the anterior-posterior axis. These inhibit Nodal
- *Bmp4:* secreted throughout the embryonic disc to generate morphogen gradients and determine cell differentiation
- *Chordin, Noggin:* inhibit Bmp4 in the midline to dorsalise the mesoderm and form the notochord and paraxial mesoderm
- *Fgf8:* works alongside Bmp4 to ventralise the mesoderm into intermediate and lateral aspects
- *Nodal:* secreted by the primitive streak to create the axis that defines the dorso-ventral axis
- *Lefty2, Nodal, Fgf8, 5-HT:* promote 'left-sidedness' in the embryo
- *Snail:* this transcription factor is upregulated on the right side of primitive node and promotes right-sided genes

KEY POINTS

- Definition of the body axes occurs during gastrulation.
- The hypoblast (primitive endoderm) is displaced by the definitive endoderm.
- The epiblast will ingress to form the three germ layers.
- The three-layered structure is the trilaminar disc and marks the formation of the axes.
- The notochord is mesodermal in origin.
- The ectoderm is the most dorsal layer.
- An anterior visceral endoderm defines the anterior-posterior axis.
- The mesoderm is subdivided into regions according to proximity to the notochord (paraxial, intermediate, and lateral).
- The endoderm generally forms the gut and structures which bud from it.
- Organs not connected to the gut tend to be mesodermal, as well as the musculoskeletal and vascular structures.
- The ectoderm generally forms the epidermis and sensory organs/tissues.

5 SEGMENTATION AND FOLDING

Segmentation is the formation of a structure in a linear series of repeating parts or segments. In the embryo, it establishes the *primary body map* to define where different structures will develop across the disc. This combines with folding in the cranio-caudal, lateral, and ventral directions to determine the final positions of organs and structures in the embryo.

SEGMENTATION AND SOMITES

From around day 20 to day 28 of development, the paraxial mesoderm condenses to form *somites*. These are bilateral transient condensations of paraxial mesoderm that will give rise to the repetitive structures of the axial skeleton including spine, ribs, muscle, and dermis. They develop progressively in pairs (either side of the midline), first cranially and then budding caudally, such that three to four paired somites form each day. In total, 42 to 44 pairs are created from the occiput base to the embryonic tail. In humans, those most caudal will disappear (as humans do not have tails) leaving only 37 pairs.

The first four pairs will form the occipital part of the skull; the next eight are cervical, then twelve thoracic, five lumbar, and five sacral. The ventral aspect of the somite becomes the *sclerotome*, which forms the vertebrae; the dorsal part forms the *dermomyotome*, which becomes the muscular, vascular, and dermal structures. The somite differentiates in response to ventral signals from expression of the *Sonic hedgehog* (*SHH*) and Noggin genes, while the dorsal aspects respond to *Wnt* and bone morphogenetic protein 4 (Bmp4) signals. A balance of these signals in the dermomyotome will activate the *myogenic factor 5* (*MYF5*) gene to form back muscle structures, the *myogenic differentiation* (*MYOD*) gene for axial and limb muscles, and the *neurotrophin 3* (*NT3*) gene for the dermis. These signals and genes are upregulated by the notochord and neural tube.

One may wonder how eight cervical somites can create seven cervical vertebrae. This occurs due a process known as *somite polarisation* and *re-segmentation*. The cranial

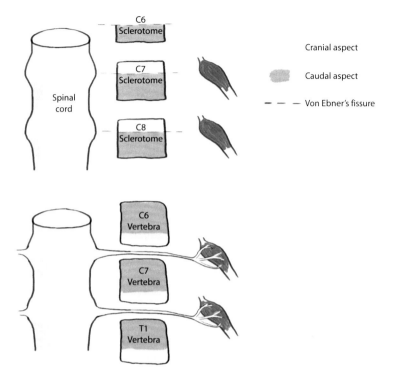

Figure 5.1 Nerves passing through Von Ebner's fissure between somite halves to reach their muscle groups (the myotome).

and caudal parts of the sclerotomal somite express different genes resulting in unequal densities between the two halves. At the divide between the cranial/caudal halves, a separation forms – *Von Ebner's fissure* – through which the nerve root will pass (Figure 5.1). The cranial section of each somite joins the caudal section above to form the vertebrae. As such, the eighth cervical somite will split, with the cranial aspect joining the caudal half of the seventh cervical somite to form the seventh cervical vertebra, and the caudal aspect of the eighth cervical somite joining the cranial part of the first thoracic somite to form the first thoracic vertebra.

FOLDING

This begins around day 22 of development and transforms the trilaminar disc into a 'tube within a tube' structure. The dorsal aspect of the disc is stiffened by the ongoing process of *neurulation*, while the yolk sac is positioned ventrally. This leads to rigidity of the disc dorsally and anchoring ventrally, so that when folding occurs it does so ventrally. The folding begins in the cranial portions and is completed caudally by day 28.

When the embryo is a disc, the most cranial structure is the *septum transversum*, followed by the *cardiogenic area* and the *oropharyngeal membrane*. These structures correspond to the presumptive diaphragm, heart, and mouth, respectively, thus making the presumptive diaphragm the most cranial structure in the disc. Folding repositions the septum transversum and cardiogenic area into the chest area, leaving the most cranial point as the oropharyngeal membrane; it also places the septum transversum caudal to the cardiogenic area so that the diaphragm sits below the heart.

After folding has completed, the embryo is covered by ectoderm everywhere except at the umbilicus, where the cord connects to the placenta. Internally, the body has divided into thoracic and abdominal cavities.

HOX GENES AND SEGMENT IDENTITY

Somite boundaries are determined by the combination and patterns of HOX gene expression. These are numerical and sequential genes that are expressed together in a code-like manner to determine the location and differentiation of each somite and organ.

CLINICAL SIGNIFICANCE

❚ Dermatomes and Myotomes

Each spinal nerve that passes through the somite will supply an area of skin (dermatome); the exception is the C1 nerve for which there is no corresponding dermatome. The nerve will also supply the myotome that is generated from the corresponding somite, so that each spinal nerve supplies a muscle or group of muscles. The dermatomes can be mapped out (Figure 5.2). While their distribution may be confusing at first, it is helpful to consider the bipedal human as a four-legged vertebrate, so that each dermatome acts as a slice.

A good knowledge of the dermatomes allows clinicians to identify the likely lesion in neurological, vascular, traumatic, and musculosketal disease/injuries. The commonly assessed dermatomes are:

- *C5:* lateral upper arm in 'regimental badge' area
- *C6:* palmar/volar aspect of thumb
- *C7:* palmar/volar aspect of middle finger
- *C8:* palmar/volar aspect of little finger
- *T4:* at the level of the nipples
- *T10:* at the level of the umbilicus
- *L3:* at the level of the knee
- *L5:* great toe
- *S1:* little toe
- *S4/5:* perianal area and anus

❚ Herpes Zoster Syndrome

This is a skin condition, also known as shingles, where a vesicular rash appears in a dermatomal distribution. Clinically, it is recognised by a rash that does not cross the midline and is isolated to a dermatome. The patient may report feeling as if they have been stung by nettles or poison ivy prior to the appearance of any skin changes. It is caused by the reactivation of *Varicella zoster* virus. The patient will have undergone an initial infection (which may have presented as chickenpox in their youth) following which the virus lay inactive in the dorsal root ganglia. Reactivation of the virus may occur due to the immunocompromise associated with another inflammatory/infective condition or physical/mental stress. The virus travels down the nerve body (of the corresponding nerve root) to create a contained immune response in the dermatome. As the virus replicates and propagates within the nerve, this causes the neuropathic pain/itching, and then generates the rash as

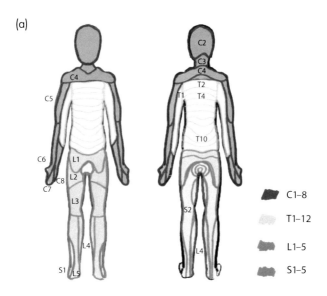

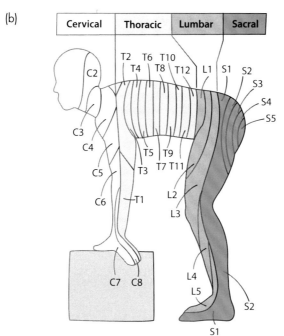

Figure 5.2 The dermatomal distribution of spinal nerves for a human as (a) bipedal and (b) on all fours.

it reaches the dermatome. It can be managed with acyclovir (antiviral drug) and prednisolone (anti-inflammatory/steroid drug).

Body Cavities

Body cavities play a crucial role in the structural organisation of the human body. The segregation into distinct cavities – such as the cranial, thoracic, and abdominal – permits the maintenance of unique pressure gradients and conditions suited to the specific functions of each area, such as creating negative pressure for the process of inhalation. From a clinical perspective, these separate compartments help limit the spread of infections and confine injuries to specific regions, thereby preventing the impairment of overall organ functionality. For instance, an abdominal penetration injury can initially localise the risk of infection and damage, without directly affecting the functionality of the lungs.

RELEVANT MOLECULES

- *Shh* and *Noggin:* morphogens responsible for ventralising the somite to form the sclerotome
- *Wnt* and *Bmp4:* morphogens that dorsalise the somite into the dermatomyotome
- *MYF5:* gene that generates back muscles from the somite
- *MYOD:* gene that codes for axial and limb formation from somites
- *NT3:* gene responsible for dermis formation
- *HOX:* sequential genes that, when expressed in certain patterns and combinations, instruct the location and outcome of somites

KEY POINTS

- Humans have 37 pairs of somites.
- There are eight cervical somite pairs but only seven cervical vertebrae.
- Somites will form the dermis (the ectoderm forms the epidermis).
- Each spinal nerve root innervates the dermis and muscle group of the somite that it penetrates; clinically this results in the dermatomal map and muscle groups.
- Clinically, the dermatomes are very important in localising pathology.
- Folding is essential for bringing the presumptive diaphragm (septum transversum) and presumptive heart (cardiogenic area) into the thorax.

6 NEURULATION AND BRAIN DEVELOPMENT

Fetal brain development is a complex and dynamic process that involves several key stages. During this period, the brain undergoes dramatic changes that lay the groundwork for all future neurological function. This chapter will delve into the major phases of fetal brain development, emphasising the transformation from simple structures to a complex organ capable of sophisticated functions. We will first look at the formation of the neural plate from the neuroectoderm, then understand how this simple structure becomes the complex brain.

NEURULATION

Neurulation is a pivotal process in early embryonic development, marking the formation of the central nervous system (CNS). It occurs after the neural plate, a flat layer of ectodermal cells, forms along the back of the embryo. The neural tube is formed during the third and fourth weeks of development from ectoderm and will produce the brain, pituitary gland, spinal cord, motor neurons, and retina. Its genesis occurs following four processes: formation, shaping, folding, and closure (Figure 6.1).

FORMATION

Ectodermal cells elongate and thicken in response to fibroblast growth factor (Fgf) signals from the notochord to form the *neuroectoderm*; the cells in this single layer structure adhere to one another with *N-cadherin* homodimers ('N' for neural). The Fgf inhibits bone morphogenetic protein 4 (Bmp4) activity that would otherwise generate epidermis from the ectodermal cells, which are held together with E-cadherin homodimers ('E' for epithelial).

SHAPING

The shaping of the neural plate generates a form that is reminiscent of a lollipop, with a broader cranial segment (the future brain) and a thinner caudal portion

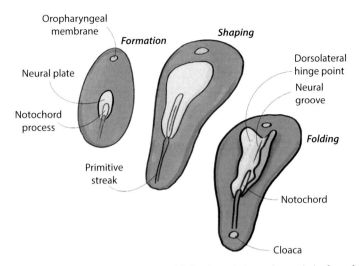

Figure 6.1 The neuroectoderm is established and then shaped, before folding to form the closed neural tube.

(the future spinal cord). This results from varying rates of convergent–extension movements. Specifically, cells in the caudal region elongate more rapidly. This differential growth is driven by distinct concentrations of morphogens in each region, with retinoic acid present in higher levels in the cranial area, and Wnt3a found in greater concentrations in the caudal section.

FOLDING

Having set out the two-dimensional structure of the CNS, the neural plate begins folding. First, the lateral edges rise, marking the initiation of the folding process. This folding occurs around a central axis lying above the notochord within a fold referred to as the *neural groove* or the *median hinge point* (Figure 6.2). The question of how a structure that varies in width along its length manages to close uniformly is answered by the presence of additional hinge points in the wider cranial segment, known as the *dorsolateral hinge points*. These additional points ensure that the broader segments align and meet at the midline simultaneously with the narrower sections. This ensures that the neural tube can form correctly and uniformly, without leaving any deficits within the tube.

CLOSURE

There are different theories as to how this occurs. The most popular is that closure first starts in the midline and then progresses in the cranial and caudal directions.

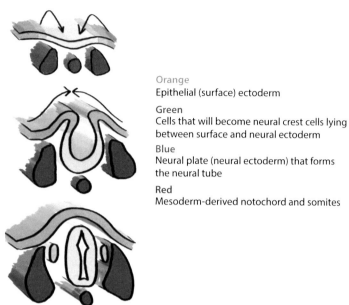

Figure 6.2 Closure of the neural tube over the central notochord.

Research in other animals, such as chickens, suggests that it may actually occur simultaneously at various points along the neural fold. Where the neural tube meets in the middle (Figure 6.2), expression of the homodimers N-cadherin and E-cadherin ensures that equivalent cells join together and separate the neuroectoderm from the surface ectoderm.

NEURAL CREST CELLS

Cells lying between the neural tube and the epidermal ectoderm will form *neural crest cells* due to an intermediate gradient of Fgf and Bmp4. These multipotent cells migrate to form many important structures around the body (listed in the 'Clinical Significance' section).

BRAIN VESICLES

The process of creating the brain vesicles begins as the cranial part of the neural tube expands and divides to form the *primary brain vesicles* (Figure 6.3). During the fourth to fifth week of development, these vesicles emerge as the *prosencephalon* (forebrain), *mesencephalon* (midbrain), and *rhombencephalon* (hindbrain). This initial division sets the stage for the complex organisation of the brain,

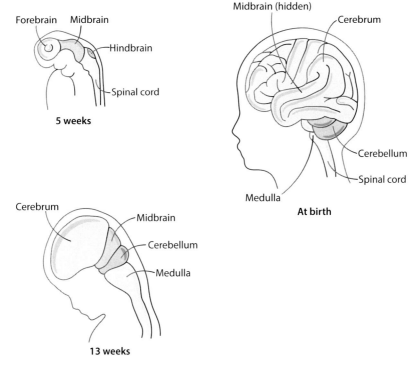

Figure 6.3 Development of the fetal brain.

delineating the early regions that will further specialise to form the adult brain's diverse structures.

As development progresses into the fifth and sixth weeks, the primary vesicles undergo further subdivision, marking the transition to the *secondary brain vesicles* stage (Figure 6.3). The prosencephalon differentiates into the *telencephalon*, which will give rise to the cerebral hemispheres, and the *diencephalon*, destined to become structures such as the thalamus and hypothalamus. Meanwhile, the rhombencephalon divides into the *metencephalon* and *myelencephalon*. The metencephalon will develop into the pons and cerebellum, which are crucial for motor control and coordination, while the myelencephalon forms the medulla oblongata, which is responsible for regulating vital body functions such as breathing and heart rate.

NEUROGENESIS

Neurogenesis is the generation of new neurons from neural stem cells, occurring most intensively between the sixth and twentieth weeks of gestation. During this period, billions of neurons are produced in the proliferative zones of the brain, notably the *ventricular zone* lining the neural tube. These newly formed neurons are destined to migrate to different parts of the brain, where they will integrate into growing neural networks. Neurogenesis is regulated by a complex interplay of genetic factors (including the transcription factors *Neurogenin* and *NeuroD*) and extracellular signal proteins (such as *brain-derived neurotrophic factor* [BDNF] and *nerve growth factor* [NGF]), ensuring the production of a diverse array of neuron types necessary for the myriad functions of the brain. This phase lays the foundation for all subsequent brain development, establishing the basic neuronal population required for the formation of functional circuits.

NEURONAL MIGRATION

Following neurogenesis, *neuronal migration*, occurring predominantly between the eighth and fifteenth weeks of gestation, is the process by which neurons move from their origin to their final destinations within the brain. This movement is essential for the proper organisation of the brain into layers and regions, each with specific functions. Neurons use various mechanisms to migrate, including *glial-guided migration*, where they travel along *radial glial cells* acting as scaffolds. Radial glial cells are present in the developing CNS of all vertebrates and are responsible for producing all neurons in the cerebral cortex, as well as specific glia (such as oligodendrocytes and astrocytes). These radial glial cells are typically found in the ventricular zone within the newly formed brain vesicles. Their structure is characterised by their long processes that extend across layers of the brain. Their regulation is managed by Fgf and Notch signalling.

AXON AND DENDRITE DEVELOPMENT

Concurrent with and after neuronal migration, neurons embark on the crucial phase of axon and dendrite development, which persists into the postnatal period. This phase is pivotal for establishing the brain's neural connections. Axons, projecting from the neuron body, navigate the nervous system to form synaptic connections with target neurons, facilitated by chemical cues and signalling molecules. Dendrites, branching extensively from the neuron, receive these synaptic inputs, significantly enhancing a neuron's signal integration capacity. The junction between axons and dendrites culminates in synapse formation, enabling neuronal communication through these specialised junctions. Synaptogenesis involves the precise alignment of axon terminals with dendritic spines, ensuring effective synaptic connectivity. This period of axonal and dendritic growth not only underscores the brain's developmental complexity but also its adaptability, allowing for the ongoing refinement of neural circuits in response to environmental changes and learning processes.

MYELINATION

Myelination begins in the late fetal period and continues into adolescence, involving the wrapping of axons with myelin, a fatty substance that insulates neural connections and significantly enhances the speed and efficiency of electrical signal transmission between neurons. Oligodendrocytes in the CNS produce myelin; this process is crucial for the functional maturation of the brain. Myelination follows a specific sequence, starting with sensory pathways, then motor pathways, and finally the association areas that are involved in complex cognitive functions. This sequential myelination is critical for the developmental milestones observed in infancy and childhood, such as walking, talking, and problem-solving.

CLINICAL SIGNIFICANCE

▌ Neural Crest Cell Derivatives

A topic commonly assessed in examinations is the derivation of neural crest cells in the fetus. It is important to remember that these are ectoderm-derived structures. A useful mnemonic is MOTEL PASS:

- **M**elanocytes and myenteric plexus
- **O**dontoblasts
- **T**racheal cartilage
- **E**ndocardial cushions and enterochromaffin cells
- **L**aryngeal cartilage
- **P**arafollicular C cells and parasympathetic nervous system post-ganglionic neurons
- **A**drenal medulla
- **S**chwann cells
- **S**pinal meninges (pia and arachnoid)

▌ Neural Tube Defects

Neural tube defects (NTDs) are types of *spinal dysraphism*: a defect that occurs during formation or closure of the spine, spinal cord, or nerve roots. In its extreme, a total dysraphism of the brain known as *anencephaly* results in a normal spinal cord and absent brain. More typically, however, pathologies occur due to varying degrees of spinal closure – *localised dysraphism*. They can be subcategorised into open and closed NTDs, denoting whether or not the spinal cord is exposed.

The most common type of NTD is *spina bifida occulta*. This is a closed NTD that occurs in 5–10% of the population where there is a small defect in the spine (e.g. missing vertebral process) that is covered by overlying tissue (Figure 6.4). It usually affects the lower spine and people will have no symptoms. It is recognised by a dimple, spot, hairy patch, or swelling in the midline of the back at the point of the gap.

The other two forms of closed NTD are *lipomyelomeningocele* and *diastematomyelia*. These occur due to a fatty lump or piece of bone in the spinal cord, respectively. The symptoms are associated with the meninges and spinal cord being trapped within these structures. This leads to changes in sensation, bladder/bowel issues, back pain, or pain on movement. It is diagnosed using magnetic resonance imaging as there may not be a visible skin lesion and clinical assessment is needed. In the infant, it may be recognised by delayed toilet training, turning of the feet (talipes), or an unusual gait.

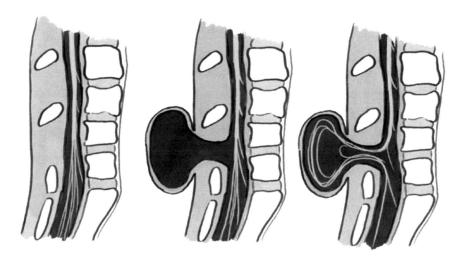

Figure 6.4 Depictions of neural tube defects. *Left:* Spina bifida occulta with a missing portion of the vertebrae, leaving only soft tissue between the cord and skin. *Centre:* Meningocele without the spinal cord. *Right:* Myelomeningocele containing the spinal cord.

There are two forms of open NTD: *meningoceles* and *myelomeningoceles*, reflecting the contents of the cavity (Figure 6.4). In meningoceles, the opening is slightly larger than in spina bifida occulta and the meninges overlying the spinal cord pass through a space between the vertebrae in a cerebrospinal fluid-filled sac. Recall that *cele* means 'cavity' and the *meninges* are the covering layers of the spinal cord. The spinal cord is not in the protruding sac; as such, cord development and function are often not affected. In a few patients, the spinal cord may tether to the sac, leading to symptoms associated with the tethered nerve root or below the lesion. Externally, meningoceles are visible as a red or purple sac.

Myelomeningoceles are the most severe form of NTD. *Myelo* is a combination of the terms meaning 'marrow' and 'of the spinal cord', reflecting the myelinated tissues. In these, the spinal cord protrudes through the opening and may be accompanied by the meninges. In the uterus, the exposed nerves can be damaged by the amniotic fluid and this, combined with the structural disconfiguration, lead to symptoms in the newborn. Again, the resultant symptoms occur at the level of, or below, the sac. The higher up the lesion, the more significant the neurological deficit.

Deficiencies in maternal *folate* can lead to NTDs. As such, pregnant patients are advised to take a daily (400 micrograms) supplement of folic acid. This dose is increased to 5 milligrams if there are any siblings born with NTDs or there is a wider a family history of the condition.

▌ Congenital Infections

Infections during pregnancy can cross the placental barrier and affect fetal brain development, leading to a spectrum of neurological and developmental disorders. Examples include:

- *Cytomegalovirus:* This is the most common congenital viral infection, and can cause microcephaly, developmental delays, and hearing loss.
- *Zika virus:* Associated with microcephaly and other severe fetal brain defects, the Zika virus highlights the impact of emerging infectious diseases on neurodevelopment.

▌ Environmental Factors

Exposure to harmful substances during pregnancy can disrupt fetal brain development, resulting in a range of cognitive, behavioural, and physical impairments. Examples include:

- *Alcohol:* Fetal alcohol spectrum disorders (FASDs) are a group of conditions arising from alcohol exposure during pregnancy, leading to lifelong learning disabilities, behavioural problems, and physical abnormalities. The typical facial features of children with FASD are microcephaly, small eye openings, low nasal bridge, smooth philtrum, small mid-face/nose, and thin upper lip (Figure 6.5).
- *Toxins:* Lead, mercury, and certain pesticides have been linked to neurodevelopmental disorders and cognitive impairments in children.

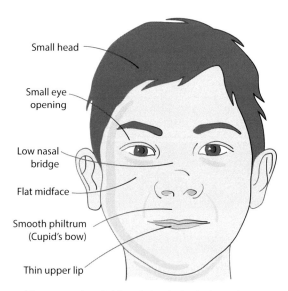

Figure 6.5 Facial features of a child with fetal alcohol syndrome.

RELEVANT MOLECULES

- *N-cadherin:* the homodimeric molecule that is present between neuroectoderm cells
- *E-cadherin:* the homodimeric molecule that is present between epidermal ectoderm cells
- *Bmp4:* induces the formation of epidermal ectoderm
- *Fgf:* inhibits Bmp4 to promote the formation of the neural plate
- *Retinoic acid:* expressed cranially to induce the shaping and structures of the neural plate
- *Wnt3a:* expressed caudally to induce the shaping and structures of the neural plate
- *Folate:* a maternal deficit in this substance leads to a higher incidence of NTDs
- *Neurogenin* and *NeuroD:* transcription factors that promote neurogenesis
- *BDNF* and *NGF:* extracellular signal proteins that manage cells in neurogenesis
- *FGF* and *Notch:* extracellular signals from these genes and their receptors manage neuronal migration

KEY POINTS

- The neural tube is derived from the ectoderm.
- Neurulation occurs in four stages: formation, shaping, folding, and closure.
- Neural crest cells form from the ectoderm and develop into many key structures.
- Spina bifida occulta occurs in up to 10% of the population.
- Maternal folic acid deficiency leads to an increased incidence of NTDs.
- Early brain development involves the formation of primary brain vesicles (prosencephalon, mesencephalon, and rhombencephalon), which then subdivide into secondary vesicles, laying the groundwork for the brain's major regions.
- Neurogenesis involves the proliferation of neural stem cells and their differentiation into neurons.
- Neurons migrate to their final destinations within the brain.
- Starting late in the fetal period and continuing postnatally, myelination enhances signal transmission speed and efficiency across neural pathways.

7 CRANIOFACIAL DEVELOPMENT

The human face forms between weeks 4 and 12 of development, initially with external structures and then the development of the intricate internal anatomy. In order to grasp the complex embryology occurring, it is useful to separate craniofacial development into distinct parts: external face, skull, palate, tongue, and pharynx. The neural crest cells and pharyngeal arches are central to the creation of many of these structures.

The *pharyngeal apparatus* consists of the *arches*, *clefts*, and *pouches*. The pharyngeal arches are paired structures consisting of mesoderm and neural crest cells (Figure 7.1). There are five arches numbered 1 to 4 and 6 (the fifth exists transiently and disappears). Between each arch lies a pouch inside and a cleft outside; this means there are only four pharyngeal pouches and clefts. The inner pouch is lined by the endoderm (similar to other internal structures), and the cleft is lined by the ectoderm (in contrast to external surfaces).

EXTERNAL FACE DEVELOPMENT

The face is formed from five *prominences*: one *frontonasal*, two *maxillary*, and two *mandibular* (Figure 7.2). These are mesenchymal proliferations of neural crest cells with the maxillary and mandibular prominences forming from part of the first pharyngeal arch. The space between the maxillary prominences is the *oral opening* and is covered by the oropharyngeal membrane – this is known as the *stomodeum*. Within the ventrolateral aspect of the frontonasal prominence, two *nasal placodes* invaginate to form *nasal pits* with medial and lateral aspects. The expansion of the maxillary prominences centres the nasal pits and creates a *nasolacrimal groove* between the presumptive maxilla and nose. This will be obliterated, leaving behind the nasolacrimal duct and lacrimal sac.

The maxillary prominences continue to grow and bring the medial nasal prominences together to meet in the midline to form the *philtrum* (Figure 7.2).

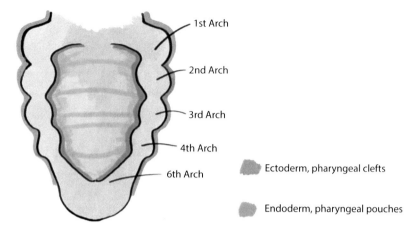

Figure 7.1 The pharyngeal apparatus consisting of arches, pouches, and clefts.

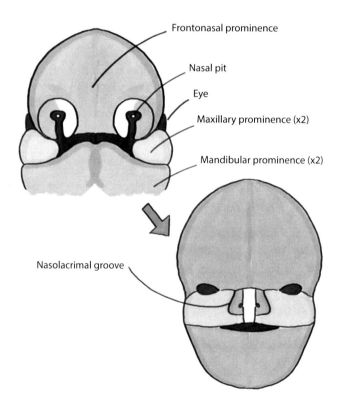

Figure 7.2 The facial prominences and their reorganisation.

This then fuses with the maxillary prominences to complete the upper lip. The final structures of each prominence are listed below:

- *Frontonasal*
    - Forehead
    - Nasal bridge
    - Midline of nose (medial)
    - Philtrum and part of upper lip
    - Alae of nose (lateral)
- *Maxillary prominence*
    - Cheeks
    - Lateral upper lip (not the philtrum)
- *Mandibular prominence*
    - Lower lip
    - Jaw

SKULL DEVELOPMENT

The germ layer origin of the bones of the skull has been a historically challenging subject. There has been dispute over which bones are ectodermal in origin, and which result from mesodermal cells. As stated previously in Chapter 5, the somites (paraxial mesoderm) contribute to the occiput. More specifically, they form the posterior bones of the cranial vault (the parietal and occipital bones) and those in the floor of the cranial fossae (the cribriform plate and *petrous* segment of the temporal bones).

The neural crest cells form the anterior bones of the facial skeleton (the *viscerocranium*); this includes the frontal, nasal, lacrimal, maxilla, volar, mandible, sphenoid, and zygomatic bones. They also form internal complex structures, such as the inferior nasal conchae, ossicles (incus, malleus, stapes) and palatine bone (hard palate), as well as the *squamous* segment of the temporal bone.

The bones of the skull develop by one of two processes. The flatter bones that create the skull vault (*neurocranium*) form through *intramembranous ossification* – where bones develop from sheets of mesenchymal connective tissue – whereas those on the floor with complex shapes (*chondrocranium*) form through *endochondral ossification*, a process by which hyaline cartilage is gradually replaced by bone (*chondro* meaning 'cartilage').

PALATE DEVELOPMENT

Palate development and fusion of the palatine shelves is critical for separating the nasal and oral cavities, which is essential for proper feeding and speech. The palate consists of a softer anterior section and harder bony palate posteriorly (Figure 7.3). It begins to form shortly after the face, from weeks 6 to 12 of gestation. For the soft segment, a *primary palate* forms as the result of fusion of the medial nasal prominences and the anterior segment of the maxillary prominence. Then a secondary palate develops from bilateral extensions of the maxillary prominences, called the *palatine shelves*, that meet in the midline and fuse.

TONGUE DEVELOPMENT

The anterior two-thirds and posterior one-third of the tongue form from different sources. The anterior segment is derived from the first pharyngeal arch, while the posterior segment develops from swellings in the third and fourth arches (Note: there is *no* contribution from the second arch). The point of fusion between anterior and posterior segments is marked by the *sulcus terminalis*; in the adult, this is bordered anteriorly by the *circumvallate papillae* (large and round taste buds). As a result of their separate develpoment, the anterior two-thirds and posterior third of the tongue have different somatic sensory and special sensory (taste) innervations.

The hypoglossal nerve provides motor function to the entire tongue. However, the anterior tongue receives sensory innervation from the *lingual branch* of the *trigeminal*

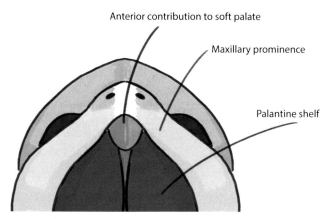

Figure 7.3 The soft (primary) palate (orange) forms as a result of contributions from the maxillary prominences (blue) and nasal prominences (green). The palatine shelves (red) are shown extending from the maxillary prominences to meet in the midline and form the hard (secondary) palate.

nerve and taste from the *chorda tympani branch* of the *facial nerve*. Meanwhile, the posterior tongue has sensory innervation from the *glossopharyngeal* and *vagal* nerves and taste from the glossopharyngeal nerve (with a small contribution from the vagus).

The intrinsic and extrinsic muscles of the tongue originate from the paraxial mesoderm that is accompanied by CN XII (hypoglossal nerve).

THE PHARYNX AND PHARYNGEAL APPARATUS

The pharynx is an intricate structure made of multiple parts derived from the parts of the pharyngeal apparatus (arch, cleft, and pouch). In embryology, the term 'pharyngeal' is often interchanged with 'branchial'. As stated previously, the arches are numbered, and this corresponds roughly with the anatomical level of contribution (cranial to caudal). Each arch is associated with a cranial nerve (CN), artery, and cartilage. Clefts are on the outside of the arches, while pouches lie on the inside.

▮ Pharyngeal clefts

The four clefts lie between the five pharyngeal arches. The first cleft is the only one to form a definitive structure in the embryo: the *external auditory meatus*. Recall that the clefts are ectodermal, so it is logical for them to form the external auditory meatus. When the second pharyngeal arch expands, it overgrows the second, third, and fourth clefts, smothering and then obliterating them. If this obliteration is incomplete then *cervical (branchial) sinuses* remain. (Note: remember the cleft lies on the outside surface, so incomplete obliteration would leave a gap; this is discussed in the 'Clinical Significance' section later.)

▮ Pharyngeal pouches

The pouches are endodermal in origin, and all four form internal structures. Similar to the first cleft, the first pouch will form the *Eustachian tube* and *middle ear cavity*. The second pouch forms the *palatine tonsils* (which are anatomically inferior to the eustachian tube). The third pouch forms the *inferior parathyroid glands* and the *thymus*. The fourth pouch will form the *superior parathyroid glands* and the *parafollicular (C) cells* of the thyroid. The third and fourth pouches are commonly assessed in examinations as candidates often mistake their inferior/superior titles.

▮ Pharyngeal arches

There is no simple way to recall the long list of structures derived from each pharyngeal arch. It is helpful to recognise patterns and the progressive anatomy in cranio-caudal sequences to try and form some order. Each pharyngeal arch will generate arteries, muscles, bone, and/or cartilage.

The nerves in the five arches (in ascending order) are the CN V (first), VII (second), IX (third), superior laryngeal branch of X (fourth), and recurrent laryngeal of X (sixth). The arteries are the maxillary (first), stapedial (second), common and internal carotids (third), subclavian and arch of the aorta (fourth), and the ductus arteriosus and pulmonary arteries (sixth). If you are able to recall the function of the cranial nerves then it may make the structures easier to remember, particularly when remembering that the facial nerve supplies the muscles of facial expression.

The first pharyngeal arch is the most cranial and is known as the mandibular arch. It consists of the maxillary and mandibular prominences and forms:

- *Artery:* maxillary
- *Nerve:* CN V (trigeminal)
- *Muscle:* muscles of mastication, mylohyoid, anterior belly of digastric, tensor veli palatini, and tensor tympani
- *Bone:* maxilla, zygomatic, squamous temporal, palatine, vomer, mandible, incus, and malleus
- *Cartilage:* Meckel's

Recall that the first cleft and pouch form other auditory structures, which align with the ossicles and muscles listed here. Also note how these are muscles which attach to the bones that they accompany (e.g. mylohyoid to the mandible). The *tensor veli* is the only muscle of the soft palate to be made by this arch (the rest are from the fourth pharyngeal arch), so this is a regularly assessed fact. *Meckel's cartilage* forms as a product of this arch, and is the precursor for endochondral ossification for the incus and malleus of the middle ear.

The second pharyngeal arch is known as the hyoid arch. Its derivatives are:

- *Artery:* stapedial (to supply the muscles of the sole ossicle in this arch)
- *Nerve:* CN VII (facial)
- *Muscle:* muscles of facial expression, posterior belly of digastric, stylohyoid muscle, and stapedius
- *Bone:* upper half of hyoid body, lesser horn of hyoid, styloid process, and stapes
- *Cartilage:* none

Although it does not give rise to any specific cartilaginous structures, you may see the second arch being referred to as *Reichart's cartilage*; this is the precursor to the bony structures that form by endochondral ossification.

The third pharyngeal arch is unnamed and forms fewer musculoskeletal structures:

- *Artery:* common carotid and proximal internal carotid
- *Nerve:* CN IX (glossopharyngeal)
- *Muscle:* stylopharyngeus
- *Bone:* greater horn and lower half of the body of hyoid
- *Cartilage:* none

Since it only forms a single muscle, and the stylopharyngeus is the only muscle of the pharynx not formed by the fourth arch, the third pharyngeal arch is commonly assessed in multiple choice question papers.

The fourth pharyngeal arch contributes significantly to the muscles and cartilage of the pharynx . It is also the first arch in which there is an asymmetry between the products of the two sides. It forms the following:

- *Artery:* proximal subclavian (right) and arch of aorta (left)
- *Nerve:* CN X (superior laryngeal branch of the vagus)
- *Muscle:* those of the soft palate [except tensor veli palatini] (levator veli palatini, palatopharyngeus, palatoglossus, musculus uvulae) and those of the pharynx [except stylopharyngeus] (superior, middle, and inferior constrictors, and the salpingopharyngeus), cricothyroid, cricopharyngeus
- *Bone:* none
- *Cartilage:* thyroid, cricoid, arytenoid, corniculate, cuneiform

Examiners often ask about the cricothyroid as it is the only intrinsic laryngeal muscle not generated from the sixth pharyngeal arch and, as a result, is innervated by the superior laryngeal nerve (rather than the recurrent laryngeal). It is also commonly assessed because it is the only tensor of the vocal cords, and this has significant clinical consequences (discussed later).

The sixth pharyngeal arch extends most inferiorly and the anatomy of its products make them easier to recollect:

- *Artery:* ductus arteriosus and pulmonary
- *Nerve:* CN X (recurrent laryngeal branch of the vagus)
- *Muscle:* intrinsic laryngeal muscles [except cricothyroid] and skeletal muscle of the (upper) oesophagus
- *Bone:* none
- *Cartilage:* thyroid, cricoid, arytenoid, corniculate, cuneiform

It is important to note that, while the fourth and sixth arches form different vascular and muscular structures, they both contribute towards the laryngeal cartilages. Recall that the recurrent laryngeal nerve loops under the ductus arteriosus on the left to rise cranially again, in order to supply the intrinsic laryngeal muscles.

The only muscles that do not form from the pharyngeal arches or somites are the trapezius and sternocleidomastoid muscles, which are products of the lateral plate mesoderm that lies beside the first four somites. This mesoderm is accompanied by CN XI (accessory nerve).

EYE DEVELOPMENT

The eye forms from ectodermal and mesodermal sources. The outer surface of the eye is formed from the ectoderm, specifically the lens, corneal epithelium, and eyelid. Neural crest cells form the sclera, corneal endothelium, corneal basement membrane, and cartilaginous components of the orbit. Somitic paraxial mesoderm contributes to the sclera, stroma, choroid, corneal endothelium, intra- and extra-ocular muscles, vessels, and the vitreous. The retina and optic nerves (and the epithelial lining of the ciliary body and iris) are formed from the neuroectoderm in a process that follows closure of the neural tube.

As the neural tube closes, outpockets of neuroectoderm – the *optic vesicles* – grow cranially to induce *lens placodes* in the surface ectoderm. These placodes develop first into pits and then vesicles (note how similar this process is to formation of the nose). Meanwhile, the optic vesicle develops into the *optic cup* – the precursor for the definitive globe.

The development of the eye is dependent on the *PAX6* gene, which modifies the tissue's responsiveness to nearby morphogens bone morphogenetic protein 4 (BMP4) and Sonic hedgehog (Shh). Shh suppresses PAX6, allowing upregulation of *PAX2*, leading to division of the optic field in the neural plate into two, whereas BMP4 works with locally upregulated fibroblast growth factor (FGF) to form the optic vesicle and cup (to form the retina), and the lens placode and vesicle.

CLINICAL SIGNIFICANCE

▌ Cleft Lip and Palate

Where the facial prominences do not completely align, a gap is left between them, known as a cleft. This can occur in the lip, palate, or a combination of both. Recalling that there are multiple prominences that meet to form facial structures, the cleft lip pathologies are as below:

- *Median cleft lip:* Incomplete fusion of nasal prominences in midline, leaving a cleft in the midline of the philtrum.
- *Oblique facial cleft:* Failure of the maxillary and nasal prominences to fuse, typically unilaterally, creating a cleft from the philtrum to the medial canthus of the eye (site of nasolacrimal system).
- *Hare lip:* Failure of maxillary and medial nasal prominences to fuse bilaterally, leaving clefts at both lateral edges of the philtrum that meet the nostrils superiorly.
- *Frontonasal dysplasia:* Hyperplasia of the frontonasal prominence that leads to the widening of the gap between nasal prominences; this leads to incomplete fusion, a flattening of the nasal bridge, and displacement of the tip of the nose.

Incomplete fusion of the maxillary extensions in the secondary palate generates a cleft palate. Since the formation of the primary palate involves the maxillary and nasal prominences, incomplete fusion of the palate often extends into the lip also, hence the pathologies often being coexistent. An untreated cleft palate can lead to various issues. The baby uses its palate for sucking milk by placing the bottle or nipple against the roof of its mouth. If it cannot latch on to drink milk, it may develop a severe nutritional deficit as a result of being unable to feed. Sometimes the malnourishment is the sign that leads to a diagnosis of a cleft palate.

Clefts of the lip and palate can also delay childhood development. Palatal clefts can lead to speech impediments that may hinder speech development. Meanwhile, the atypical anatomy can lead to an increased risk of recurrent sinus and ear infections that may affect hearing, resulting in speech and interaction milestone delays.

▌ Branchial Cysts and Sinuses (Pharyngeal Cleft Pathology)

Failure of complete overgrowth and obliteration of the second to fourth clefts by the second pharyngeal arch leaves the space that these clefts were occupying. This becomes a branchial *cyst* if it is sealed from the external surface and a *sinus* if there is a tract to the skin. These cysts/sinuses can grow large (compressing surrounding structures) or become infected: both are indications for potential surgical excision.

They can occur anywhere along the anterior border of the sternocleidomastoid (SCM) in the anterior triangle of the neck. The location of the cyst/sinus along the SCM depends on which cleft has not been obliterated, such that second-cleft cysts will be located more superiorly (typically inferoposterior to the mandible), third-cleft cysts are posterolateral to the laryngeal cartilage, and fourth-cleft cysts are in the thyroid region. The third- and fourth-cleft cysts can have internal openings, which make them prone to recurrent infections.

Thyroids, Parathyroids, and Thymus (Pharyngeal Pouch Pathology)

Most commonly, absence of or anomalies in the derivatives of the third and fourth pharyngeal pouches are noticed clinically: during head and neck operations, in endocrine disorders (of calcium), or during endocrine surgery. These disorders manifest as absences or dysplasia in the parathyroid, thymic, and parafollicular tissues.

Even under normal circumstances, the parathyroid glands have the most variable anatomy of all structures in the body, which is a challenge for endocrine surgeons. When the parathyroids form with the thymus (third pharyngeal pouch) and thyroid gland (fourth pharyngeal pouch), they accompany them as they migrate from the neck into their final positions inferiorly. This leaves the fourth-pouch parathyroids superiorly in position, typically posterior to the middle of each thyroid lobe. The third-pouch inferior parathyroids often deposit just below these; however, they can continue migrating anywhere between the thyroid in the neck and the thymus behind the sternum, so their position is very variable between individuals. The location of the parathyroids near the thyroid puts them at risk during thyroidectomies; as such, it is routine post-operatively to check serum calcium and parathyroid levels to ensure the patient does not develop severe hypocalcaemia.

DiGeorge Syndrome (Third and Fourth Pharyngeal Arches)

Also known as *velocardiofacial syndrome*, this condition is commonly assessed in examinations as its pathology involves multiple systems. It occurs due to a deletion in the long arm of chromosome 22 (*22q11*). This results in a hypoplasia of the third and fourth pharyngeal arches and pouches leading to:

- Hypoplasia of the thyroid
- Thymic hypoplasia
- Hypoparathyroidism
- Heart outflow tract disorders (neural crest cells from this region contribute to the conotruncal cushions of the heart)

This hypoplasia also impacts the first and second pharyngeal arches leading to micrognathia, sensorineural hearing loss (dysplastic ossicles), conductive hearing loss (recurrent infections), and cleft palate (*velum* meaning 'palate' in Latin). For examinations, the mnemonic *CATCH-22* is a useful aide-mémoire: Cardiac anomalies, Abnormal facies, Thymic hypoplasia, Cleft palate, and Hypocalcaemia.

Laryngeal Nerve Palsy

The fourth pharyngeal arch generates both the superior laryngeal nerve and the thyroid gland. In thyroidectomies, this nerve is at risk of damage. The external branch of the superior laryngeal nerve provides motor function to the cricothyroid, while the internal branch provides sensation to the mucous membrane of the larynx. A unilateral external superior laryngeal nerve palsy is associated with changes in the pitch of the voice, while a bilateral palsy leads to hoarseness. In contrast, a unilateral palsy in the recurrent laryngeal nerve causes hoarseness, while a bilateral palsy can cause fatal dyspnoea if the cords lie in the abducted position and block the airway.

RELEVANT MOLECULES

- *PAX6:* gene responsible for initiating and regulating the complex development of the eye
- *BMP4:* promoted by PAX6 to upregulate FGF and develop the optic cup and lens of the eye
- *FGF:* works with BMP4 to locally induce cells to differentiate into optic structures
- *Shh:* promoted by PAX6; this then inhibits PAX6 to upregulate expression of PAX2, leading to division of the optic field into two (avoiding cyclopia)

KEY POINTS

- Pharyngeal arches are mesoderm.
- 'Branchial' and 'pharyngeal' are interchangeable terms.
- Clefts are on the outside surface of the pharyngeal apparatus and ectodermal in origin (cleft contains the letters 'ect' and ectoderm forms the outside layer of embryos).
- Pouches are on the inside of the apparatus and formed from endoderm.
- The somites form the flat bones of the skull vault (neurocranium) through membranous ossification.
- Endochondral ossification is used to form bones with more complex shapes.
- The tongue develops in two parts with the anterior two-thirds and posterior third having different vasculature and innervation.
- Only the first pharyngeal cleft forms a structure: the external auditory meatus.
- The third pharyngeal pouch forms the inferior parathyroid glands, while the fourth pouch forms the superior parathyroids.
- There is no fifth pharyngeal arch in humans.
- The stylopharyngeus is formed from the third pharyngeal arch and is the only muscle of the pharynx not to be formed by the fourth arch.
- The cricothyroid is the only intrinsic laryngeal muscle to be formed by the fourth arch (all others are formed from the sixth arch) and, therefore, has a different innervation to the other muscles.
- The retina is formed from neuroectoderm.

8 EAR AND NOSE DEVELOPMENT

In Chapter 7, we now understand craniofacial development from its origin in the pharyngeal arches and neural crest cells. In this section, we delve deeper into the formation of two key structures: the ear and nose. The development of these structures represents two of the most intricate processes in human embryogenesis, culminating in structures essential for key sensory functions: hearing, balance, and olfaction. These processes begin early in embryonic life and are characterised by precise genetic regulation, cellular differentiation, and morphological changes.

Central to the development of the complex structures of the face is the *Dickkopf WNT signalling pathway inhibitor 1* (*DKK1*) gene. DKK1 modulates the Wnt signalling pathway. It inhibits Wnt signalling to regulate the migration, proliferation, and differentiation of cranial neural crest cells, essential for forming the facial skeleton, cartilage, and connective tissue. Inhibition of Wnt signalling by DKK1 ensures proper patterning and morphogenesis, preventing craniofacial anomalies such as cleft palate and craniosynostosis. Additionally, DKK1 interacts with bone morphogenetic protein (BMP) and fibroblast growth factor (FGF) pathways, coordinating the complex processes of craniofacial development and ensuring balanced growth and differentiation of craniofacial tissues.

EAR DEVELOPMENT

The *outer ear* (pinna) emerges first, heralded by the appearance of six small hillocks around the first pharyngeal cleft during the sixth week of gestation (Figure 8.1). These hillocks, arising from the first and second pharyngeal arches, coalesce and expand to sculpt the auricle/pinna by the twelfth week. Parallel to this, the external auditory canal forms, carving a path through the first pharyngeal cleft to link the newly formed outer ear with the developing middle ear – preparing a pathway for sound transmission.

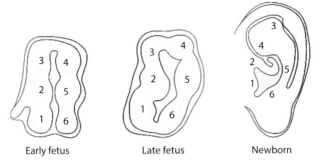

Figure 8.1 The six hillocks of the embryonic ear.

The *middle ear*'s development is characterised by the transformation of the first pharyngeal pouch into the tympanic cavity and the eustachian tube. This expansion creates a chamber that will house the ossicles – the malleus, incus, and stapes – which themselves originate from the pharyngeal arches. These ossicles, through a process of differentiation and ossification, become the vital link in conveying vibrations from the tympanic membrane to the inner ear. Their development is a delicate balance of growth and morphogenesis, ensuring their eventual role as the smallest bones in the human body, integral to the auditory process. This process is managed by many transcription factors including *T-box 1 (TBX1)*, *eyes absent homolog 1 (EYA1)*, *sine oculis homeobox homolog 1 (SIX1)*, *paired box 2 (PAX2)*, *GATA binding protein 3 (GATA3)*, and HOX. *Msh Homeobox 1 (MSX1)* is implicated in the ossification process of the middle ear ossicles and is also involved in the formation of the hard palate and nasal septum.

Concurrently, the *inner ear* begins its formation from the *otic placode*, a specialised region of ectoderm that invaginates to form the *otocyst*. Patterning of the inner ear occurs due to a gradient of Sonic hedgehog (Shh), including the dorsoventral axis. From this vesicle, the labyrinthine architecture of the cochlea and the vestibular apparatus emerges. The cochlea spirals into its hallmark snail-like shape, housing the organ of Corti, the epicentre of sound reception (Figure 8.2). Meanwhile, the vestibular system, with its semicircular canals, utricle, and saccule, matures into the body's equilibrium centre, navigating the complexities of motion and balance.

The development of the ear, from its earliest embryonic whispers to its final anatomical formation, underscores the intricate interplay of genetic, cellular, and morphological processes essential for the creation of our auditory and vestibular systems. This journey not only fascinates with its precision and complexity but also emphasises the importance of understanding embryological development in diagnosing and addressing congenital disorders of hearing and balance.

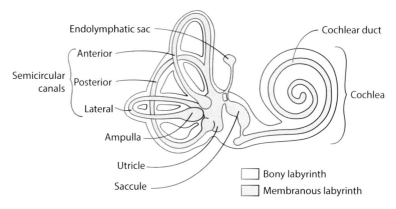

Figure 8.2 Inner ear anatomy.

NOSE DEVELOPMENT

The development of the nose is a critical aspect of facial morphogenesis, encompassing the formation of the external nose, nasal cavities, and the olfactory system. This process is intricately regulated and involves the transformation of the frontonasal prominence along with the medial and lateral nasal processes in the embryo.

As the embryo progresses into the fifth and sixth weeks of development, the frontonasal prominence gives rise to the medial and lateral nasal processes. The medial processes evolve to form the septum, crest, and tip of the nose, laying the groundwork for the nostrils and the philtrum of the upper lip. Concurrently, the lateral processes contribute to the sides of the nose, shaping its width and structure. Central to this phase is the formation of nasal placodes, thickened patches of ectoderm that mark the initial blueprint of the nasal region (Figure 8.3). These placodes deepen into nasal pits, gradually morphing into nasal sacs. This transformation is pivotal, as it establishes the *primary nasal cavities*, setting the stage for the development of the complex network of air passages and the olfactory system.

The fusion of the medial nasal processes is a hallmark event, creating the intermaxillary segment. This critical fusion not only influences the aesthetic contours of the upper lip and the base of the nose but also plays a significant role in forming the nasal septum, dividing the nasal cavity into two chambers. It is within these chambers that the olfactory epithelium emerges, a specialised tissue endowed with olfactory receptors, essential for the sense of smell. The developmental narrative extends into the formation of the nasal cavities, characterised by their division into

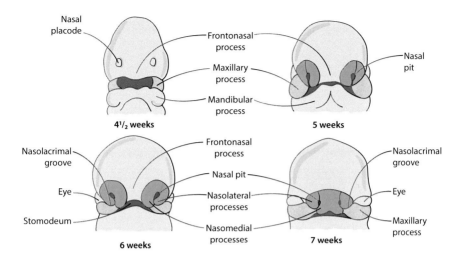

Figure 8.3 Nasal development in the embryo.

the superior, middle, and inferior meatuses. This segmentation, achieved through the growth of the nasal conchae, optimises the nasal cavities' functions in filtering, humidifying, and warming the air we breathe. Additionally, the paranasal sinuses begin to materialise as extensions of the nasal cavities. Although their development extends into adolescence, their inception during embryogenesis underscores their role in augmenting the resonance of our voices and lightening the weight of the skull.

CLINICAL SIGNIFICANCE

▌ Congenital Hearing Loss

Congenital hearing loss refers to hearing impairment present at birth, which can impact the child's ability to develop speech, language, and social skills. Causes a range from genetic factors, including mutations affecting the development and function of the auditory system, and environmental factors such as maternal infections (e.g. rubella, cytomegalovirus) and premature birth. The severity can vary from mild to profound, impacting one or both ears.

In the UK, such hearing losses are detected in the *Newborn Screening Programme* within a few weeks of birth and before discharge from hospital. Intervention strategies include social, medical, and surgical such as sign language education, hearing aids, and cochlear implants.

▌ Microtia

Microtia is a congenital malformation of the external ear, ranging from mild abnormalities to complete absence of the ear (*anotia*). It can occur as an isolated condition or as part of syndromes such as *Treacher Collins syndrome* and *Goldenhar syndrome*. The condition is unilateral in most cases but can affect both ears, potentially impacting hearing due to associated atresia (absence or closure) of the external auditory canal.

It is diagnosed with physical examination and hearing assessment. Depending on the severity and size of the defect, as well as patient preference, microtia can be managed by reconstructive surgeons using cartilage grafts from the ribs in staged procedures.

▌ Transcription Factor Pathologies

There are many conditions that can occur from dysregulation of the genes for the transcription factors responsible for development; these include:

- *TBX1:* A member of the T-box family of transcription factors, TBX1 is crucial for the development of structures derived from the pharyngeal arches and pouches. Mutations in the TBX1 gene are associated with DiGeorge syndrome, which can include middle ear malformations among its spectrum of symptoms.
- *EYA1:* This transcriptional co-activator is vital for organ development, including the middle ear. Mutations in EYA1 can lead to branchio-oto-renal (BOR) syndrome, characterised by branchial arch anomalies, hearing loss, and renal dysplasia.

- *SIX1:* Working closely with EYA1, SIX1 is involved in the development of the ear and kidney. Mutations in SIX1 are also associated with BOR syndrome, highlighting its role in ear development.
- *PAX2:* This transcription factor is important for the development of the otic vesicle and the subsequent structures of the inner ear, but it also influences middle ear development indirectly through its role in the overall development of the ear.
- *GATA3:* This gene is involved in the development of the auditory system, including the middle ear. Mutations in GATA3 can lead to hypoparathyroidism–deafness–renal dysplasia (HDR) syndrome, which includes hearing loss as a key feature.
- *HOX genes:* The HOX family of transcription factors are known for their role in patterning the body axis and limb development; they also contribute to the morphogenesis of the middle ear, particularly in the specification of structures derived from the pharyngeal arches.

Choanal Atresia

A condition where the choanae (the nasal passages connecting the nasal cavities to the pharynx) are blocked, usually by tissue, affecting breathing after birth.

Nasal Septal Deviation

A common condition where the nasal septum, the cartilage and bone dividing the nasal cavity, deviates from the midline. While it can occur due to trauma, significant deviations are often congenital. *SRY-box transcription factor 9 (SOX9)* is a key transcription factor in the development of the nasal septal cartilage as well as other cartilaginous structures in the nose and ear.

Arhinia

A very rare congenital condition characterised by the partial or complete absence of the nose at birth. It is associated with a spectrum of other developmental anomalies.

RELEVANT MOLECULES

- *FGF:* critical for initiating the development of the otic placode and the nasal placodes
- *WNT:* proteins expressed by this gene family are involved in the regulation of otic placode formation and nasal structure differentiation
- *BMP:* proteins expressed by this gene family play a role in the patterning and morphogenesis of the ear, particularly in the specification of the dorsal-ventral axis of the otic vesicle, and are involved in the development and ossification of nasal structures
- *Shh:* morphogen involved in the ventral patterning of the developing ear and in the growth and midline patterning of facial structures, including the nasal septum
- *PAX: PAX2* and *PAX8* genes are involved in early ear development, including otic placode formation, while *PAX6* plays a role in nasal development and olfactory system formation
- *HOX:* genes that contribute to the regional specification within the pharyngeal arches, influencing the development of structures contributing to the ear and parts of the nose
- *TBX1:* gene critical for the development of the pharyngeal apparatus
- *SOX9:* gene involved in the differentiation of chondrocytes, playing a key role in the development of the cartilaginous structures of the ear and the nasal septum
- *MSX1:* regulates bone development and is implicated in the ossification process of the middle ear ossicles and the formation of the hard palate and nasal septum
- *DKK1:* acts as an inhibitor of Wnt signalling, involved in the development of craniofacial structures, including the nose, and potentially influencing ear development
- *EYA1, SIX1,* and *GATA3:* genes essential for middle ear development

KEY POINTS

- Ear and nose development begin with the induction of the otic and nasal placodes, respectively, from the surface ectoderm, influenced by signals from the surrounding tissues.
- The otic placode, which forms the inner ear, invaginates to create the otic vesicle, the precursor to the cochlea and vestibular apparatus.
- Nasal placodes deepen to form nasal pits, eventually giving rise to the nasal cavities and the olfactory epithelium critical for the sense of smell.
- The first and second pharyngeal arches play a crucial role in forming the middle and outer ear structures, including the ear ossicles and the external auditory canal.
- Fusion of the lateral and medial nasal processes forms the primary palate and nasal septum, essential for the separation of the oral and nasal cavities and the formation of the upper lip and nose structure.
- Mesenchymal condensations within the pharyngeal arches differentiate into the cartilage and bone of the ear ossicles, and contribute to the structural framework of the nose.

9 HEART AND VESSEL DEVELOPMENT

The heart is the earliest functional organ to develop in the embryo, and the heartbeat can be detected in early booking or viability scans for pregnancy from as soon as 6 weeks of development (4th week of gestation). This milestone is occasionally used in arguments surrounding personhood of the fetus and has formed part of many controversial policies surrounding abortion. The heart begins pumping as a tube from day 22, and its development involves a significant remodelling to transform to a multi-chamber, highly specialised organ. The process of heart development is highly reliant on three genes working synergistically: *NKX2.5*, *GATA4*, and *TBX5*.

HEART TUBE

The early heart is tubular in shape. Its endocardium and myocardium develop from a *cardiogenic field* in the cranial third of the lateral plate mesoderm. Initially, a *primary heart field* forms and coalesces by the third week to create a *cardiac crescent* of progenitor cells. The epicardium (which also forms the coronary vessels) originates from a separate source of mesenchymal cells at a later stage in heart development. Lateral folding of the embryonic disc brings the sides of the cardiac crescent together to form a tube (Figure 9.1), while cranio-caudal and ventral folding bring the heart precursor tissue into the centre of the chest.

At this point, the heart is suspended from the body wall by the *dorsal mesocardium*, which is later obliterated to form the *transverse pericardial sinus*. In the developed heart, this lies posterior to the ascending aorta and pulmonary trunk, and anterior to the superior vena cava (SVC), separating the arterial and venous outflows of the heart. As such, it is used in cardiothoracic surgery to identify the vessels for temporary clamping.

The linear heart tube that results from folding has a caudal inflow end and a cranial outflow end. It consists of an endothelium surrounded by a contractile myocardium (Figure 9.1). This tube will elongate in order to prepare for folding.

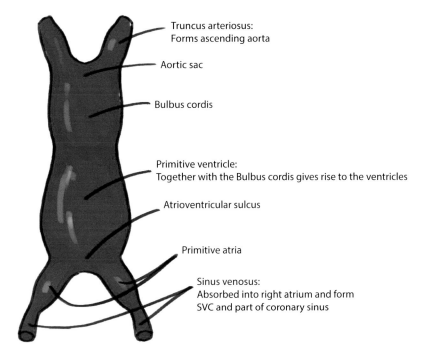

Figure 9.1 The linear heart tube and final products.

HEART FOLDING

The looping of the heart determines the definitive positions of the chambers. It first lengthens and folds into a C-shape (with the right side as the outside curvature), before looping further into an S-shape (Figure 9.2). In order to elongate, cardiac progenitor cells are recruited from a *secondary heart field* and added to the arterial and venous poles; this process displaces the structures as follows:

- *Bulbus cordis inferiorly, ventrally, and to the right:* this moves the presumptive right ventricle anteriorly in the chest
- *Primitive ventricle to the left:* this positions the presumptive left ventricle
- *Primitive atrium posteriorly and superiorly:* this places the atria superior to the presumptive ventricles

The primitive atria contribute little to the definitive atria and are incorporated into nearby structures. They remain as the *auricles* of the atria. The right atrium forms as a product of the *right sinus horn*, while the left is produced by the *pulmonary outflow vessels*. The right ventricle is predominantly formed from the *bulbus cordis*;

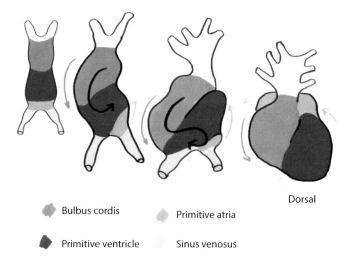

Figure 9.2 The linear heart tube folding into a 'C' and then an 'S'.

this structure also contributes to the outflow tract of the left ventricle. The bulk of the left ventricle is formed from the *primitive ventricle*.

ATRIAL SEPTATION

Septations divide the folded heart tube into four chambers. First, the atria and ventricles are separated by *endocardial cushions* growing inwards from the endocardial walls. These cushions are located dorsally and ventrally and meet in the midline to form the *septum intermedium*. Their development and position is dependent on *retinoic acid* signalling: this causes the cells to undergo *epithelial-to-mesenchymal transition* to form tissue that is different from the endocardium and myocardium. This process is often disrupted in patients with *Down's syndrome*.

Now that the heart has been divided into two compartments of atria and ventricles, the *septum primum* (Latin for 'first fence') grows from the roof of the primitive atria around day 28 to separate them into left and right sides (Figure 9.3). It is a membranous septum that grows towards the septum intermedium. The space between the two septa is known as the *ostium primum* (Latin for 'first door'). Towards the end of the sixth week of development, the septum primum fuses with the septum intermedium to obliterate the ostium primum. Simultaneously, *apoptosis* occurs at the superior margin of the septum primum to create a new gap – the *ostium secundum* ('second door'). At its completion, the *septum secundum* ('second fence') develops from the roof of the atrium on the right side of the septum

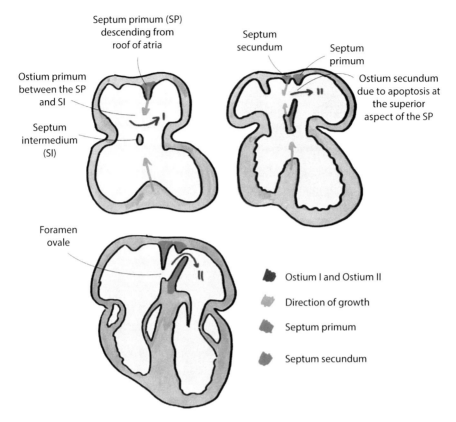

Figure 9.3 Septation of the heart.

primum and ostium secundum. In contrast to the septum primum, the secundum is thick and muscular, but it does not reach the septum intermedium. As a result, the *foramen ovale* is produced in the communication between the right and left atria through the ostium secundum.

This process is often challenging and confusing for students of embryology. It is important to consider each step in turn and leave the Latin terminology to one side. Each part of the process is named sequentially in a logical manner. Put simply, a septum begins developing from the roof of the atria (septum primum), which by forming creates a hole between the two atria by definition (ostium primum). When the septum primum reaches the septum intermedium, it takes up the space occupied by that hole. A second hole forms within the septum primum due to apoptosis (ostium secundum), and a second septum forms from the roof of the atria (septum secundum). The second septum has a gap between itself and the septum intermedium, and this gap – in combination with the ostium secundum in the septum primum – is collectively known as the foramen ovale as it is the oblique hole between the gap left by the septum secundum and the ostium secundum.

It is important to recognise that the fetal blood pressure in the right atrium is higher than in the left atrium. This is because it receives both oxygenated blood from the placenta and the deoxygenated blood from the body; meanwhile, the left atrium receives little blood from the compressed (uninflated) fetal lungs. Furthermore, the pulmonary pressure is high due to the deflated lungs, so there is an increased afterload for the right side of the heart. Therefore, the blood flows down the pressure gradient from the right atrium to the left through the foramen ovale; this also prevents the septum primum and secundum from fusing as the force of the flow pushes away the membranous septum primum. This *right-to-left shunt* is important in the embryo as it transfers oxygenated blood from the right (pulmonary) side of the circulation to the left (systemic) side of the circulation for pumping to perfuse the body.

VENTRICULAR SEPTATION

An *interventricular muscular septum* develops from the caudal aspect of the ventricle around the fourth week of development. This grows towards the septum intermedium but does not meet it. This leaves an *interventricular foramen* such that both ventricles share an outflow tract and oxygenated blood in the right ventricle reaches the systemic (left-side) circulation.

HEART VALVE FORMATION

There are four valves in the heart. The mitral (left) and tricuspid (right) divide the atria from the ventricles, and the aortic (left) and pulmonary (right) valves are located at the ventricular outflow tracts. Their development differs very slightly. While both sets originate from endocardial cushions (cardiac cells that have undergone epithelial-to-mesenchymal transition), the outflow valves require the additional ingression of neural crest cells into the cushions. The overall process is regulated by the receptor *Notch1* (and its ligand *Jagged1*).

REVERSAL OF HEART FLOW AND CLOSURE OF THE FORAMEN OVALE

The right-to-left shunting of the fetal circulatory system persists until birth in order to ensure that the systemic circulation receives maternal oxygenated blood. As the baby takes its first breath, expansion of the alveoli leads to dilatation of the pulmonary capillaries that lie on their surface. This decreases the pressure within the alveolar capillaries, leading to a drop in pulmonary pressure and generating blood flow through the lungs. This returns to the heart via the left atrium and increases the blood pressure of the left side of the heart. As a result, the raised pressure in the left atrium pushes the septum primum away and closes the foramen ovale, although it can persist in up to 25% of individuals. This process is important to stop the right-to-left shunt and avoid venous admixture once the baby is born and the maternal oxygenated blood supply stops. If this shunt remained open, then deoxygenated blood would be entering the systemic circulation from the right atrium.

FORMATION AND SPIRAL SEPTATION OF THE GREAT VESSELS

During the fifth week of development, *septation* of the outflow tract leads to closure of the interventricular foramen. This process begins when neural crest cells from the fourth and sixth pharyngeal arches migrate into the outflow tract (*truncus arteriosus*). They undergo epithelial-to-mesenchymal transition to form two *truncus cushions*. These proliferate to meet in the midline and then grow to meet the interventricular septum, forming the membranous portion of this septum.

As these cushions meet in the midline, they spiral to separate the outflow tract into the aortic and pulmonary aspects. The pharyngeal arches will then contribute to the arterial vasculature (as discussed in Chapter 7) through aortic arches (Figure 9.4). These arches will combine with a *primitive dorsal aorta* that has formed in a pair from the lateral plate mesoderm. The combined structures join in the midline to create the descending aorta. The *arch of the aorta* (not the aortic arches) is generated from the heart tube, pharyngeal aortic arches, and dorsal aorta in the following segments:

- *Initial:* from the truncus arteriosus of the heart tube
- *Ascending aorta:* the aortic sac (the distal part of the truncus arteriosus)
- *Transverse component:* from the left fourth pharyngeal aortic arch
- *Descending aorta:* dorsal aorta (mainly left component)

When considering the *pharyngeal aortic arches* and their derivatives, they are as follows:

- *First aortic arch:* maxillary artery
- *Second aortic arch:* stapedial artery
- *Third aortic arch:* common carotid and proximal internal carotid
- *Fourth aortic arch*

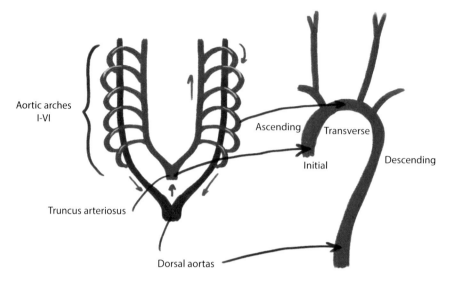

Figure 9.4 The primitive aortic structures and their corresponding parts.

- *Left:* transverse component of the anatomic arch of the aorta
- *Right:* proximal right subclavian artery
- Sixth aortic arch
  - *Left:* left pulmonary artery and the *ductus arteriosus*
  - *Right:* right pulmonary artery

The ductus arteriosus connects the trunk of the *pulmonary artery* to the *proximal descending aorta*. This is to enhance the right-to-left shunt of the blood flow in the embryo and increase the oxygenated blood in the systemic circulation. Any blood not shunted from the right to the left atrium would pass through the right ventricle into the pulmonary vessels and then be shunted into the aorta through the ductus arteriosus.

When the fetus is born, the levels of placental *prostaglandin* (PG) E2 drop and *bradykinin* is released from the ventilated lungs. This change closes the ductus arteriosus and forms the *ligamentum arteriosum* (which the left recurrent laryngeal nerve loops around).

DEVELOPMENT OF THE VENOUS SYSTEM

The inferior vena cava (IVC) develops through extensive remodelling of the venous system. Initially, the *vitelline veins* from the yolk sac and the umbilical veins from the placenta are paired structures connected through extensive plexuses connecting their right and left parts. They work in a system with the *cardinal veins* – the early venous system of the 'body' of the embryo. The *sinus venosus* (the inflow aspect of the heart tube) receives blood from the vitelline vein, umbilical vein, and the common cardinal vein.

As the liver develops and encroaches on the vitelline vessels, the left vitelline system becomes the *portal vein* and the right vitelline becomes the *suprahepatic portion of the IVC*. An anastomosis forms between the left umbilical and vitelline veins called the *ductus venosus*. This allows oxygenated and nutrient-rich umbilical blood to bypass the liver (and first-pass metabolism) and directly enter the IVC to be delivered to the heart. This remains as the *ligamentum venosum* of the liver (which separates the caudate lobe from the left lobe).

In the embryo, the early venous system is named according to the location relative to the heart with the *anterior cardinal* (*pre-cardinal*) veins draining the head, neck, upper torso, and upper limbs, while the *posterior cardinal* (*post-cardinal*) veins drain the body and lower limbs. As the embryo grows in size and the organs develop, the posterior cardinal veins are joined by the *subcardinal* and *supracardinal* veins. The anterior and posterior systems undergo significant remodelling to form veins draining into the SVC and IVC, respectively.

CLINICAL SIGNIFICANCE

Congenitial cardiovascular malformations occur in about 1% of all live births and account for about 20% of all congenital deformities.

▎ Cyanotic Babies and Duct-Dependent Circulations

A central concept to understanding the clinical embryology of the heart is why cyanosis occurs following specific heart malformations. Central cyanosis occurs when the level of deoxygenated haemoglobin is above 5 g/dL (with oxygen saturations below 85%). In the newborn, it is important to determine the timing and symptoms of the 'blue baby', as the diagnosis varies depending on the time since birth (Table 9.1). In blue babies, if the cyanosis worsens when the baby cries then the likely cause is a heart defect due to the raised blood pressure, as this increases with crying. If the cyanosis improves, the cause is likely respiratory and responding to the increased lung capacity and oxygenation.

If the baby is born pathologically cyanotic (i.e. persistently blue after the initial 5 to 10 minutes), then this means that there is difficulty in oxygenating blood. The neonate may have a persistent or large right-to-left shunt where deoxygenated blood bypasses the lungs and enters the systemic circulation through a shunt in the heart. The two most commonly assessed causes of this are *tetralogy of Fallot* (*ToF*) and *transposition of the great arteries* (*TGA*) (Table 9.1). However, it also occurs with *total anomalous pulmonary venous return, tricuspid atresia, truncus arteriosus,* and *hypoplastic left heart syndrome* (*HLHS*), sometimes referred to as the 'five Ts (and one H)'.

Table 9.1 Presentation of heart defects and possible underlying causes

Type of lesion	Left-to-right shunt	Right-to-left shunt	Common mixing	Well children with obstruction	Sick neonates with obstruction	Duct-dependent circulation
Symptoms	Breathless or asymptomatic	Blue	Breathless and blue	Asymptomatic	Collapsed with shock	Delayed until around 3 weeks
Causes	ASD VSD, PDA	ToF TGA	AVSD	AS PS, CoA	CoA HLHS	PA, AS HLHS, CoA

Abbreviations: AS – aortic stenosis; ASD – atrial septal defect; AVSD – atrio-ventricular septal defect; CoA – coarctation of the aorta; HLHS – hypoplastic left heart syndrome; PA – pulmonary atresia; PDA – patent ductus arteriosus; PS – pulmonary stenosis; TGA – transposition of the great arteries; ToF – tetralogy of Fallot; VSD – ventricular septal defect.

If the baby is born normal in colour and then becomes cyanotic in the first few days of life, this reflects a progressively cyanotic and *duct-dependent circulation*. This is a circulation that depends on the shunt of the ductus arteriosus between the pulmonary artery and the aorta to pump oxygenated blood systemically. The closure of the ductus arteriosus in these patients will reveal the life-threatening pathology: either blood is unable to be oxygenated via the lungs (e.g. pulmonary atresia), or oxygenated blood is not reaching the systemic circulation (e.g. critical aortic stenosis [AS]). If blood is unable to reach the lungs through the pulmonary vessel (e.g. pulmonary atresia), then the shunt is necessary to transport blood from the aorta into the pulmonary artery, bypassing the blockage at the outflow tract. The same bypass mechanism would work for oxygenated blood reaching the aorta to bypass the outflow blockage. These newborns require medicating with PGE1/2 to keep the duct open (recall that levels of PG decline with the removal of the placenta) while definitive surgical management is prepared.

Where the newborn is able to oxygenate the blood but there is mixing of arterial and venous blood (common mixing; venous admixture), then it will present with symptoms of severe anaemia – peripheral cyanosis and breathlessness. This most commonly occurs in atrio-ventricular septal defects (AVSDs). If the baby has signs of haemodynamic shock (e.g. low blood pressure), then one has to consider the function of the left side of the heart and its outflow. It may be that the pumping mechanism of the heart is ineffective and unable to generate sufficient pressure to perfuse the organs. For these patients, the differential includes conditions such as coarctation of the aorta (CoA) or an interrupted aortic arch.

▌ Septal Defects

These conditions occur on a very broad clinical spectrum from asymptomatic to life-threatening. They can be auscultated as *murmurs* of the heart as they lead to turbulent flow; however, it is important to remember that a larger defect may have a quieter or absent murmur as blood easily passes through the defect. Owing to the higher pressures in the left side of the heart, these typically cause a left-to-right shunt where oxygenated blood re-enters the pulmonary circulation. If the defect is sufficiently large then this can lead to signs of breathlessness, either due to insufficient oxygen delivery or an overloaded pulmonary system (pulmonary hypertension). Some patients may present in adulthood with signs of heart failure due to pathological cardiomegaly from the persistently elevated pressures.

Septal defects occur due to inadequate growth or closure of the septal processes in heart development. They often form part of a greater heart abnormality, clinical syndrome, or genetic mutation. Around 5–10% of all babies with congenital heart defects have some form of atrial septal defect (ASD), and about 33% of all congenital heart disease is due to a ventricular septal defect (VSD): approximately 3/1000 live births. For the ventricles, it is typically the membranous component of the septum that is affected.

Rarely, these may progress to *Eisenmenger's syndrome* where the patient presents with cyanosis and dyspnoea several years later (typically aged 10 to 15 years). This happens because the higher pressure in the left ventricle (or atrium) shunt blood across into the right side of the heart and cause pulmonary hypertension. The increased pressure causes hypertrophy of the pulmonary vessels and reduces their compliance, which in turn causes pathological hypertrophy of the muscles of the right side of the heart. When the pressure in the right chambers rises sufficiently, the shunt reverses and blood passes from the right to the left. This leads to deoxygenated blood entering the systemic circulation and the patient becomes cyanotic as a result.

▌ Transposition of the Great Arteries and Outflow Tract Pathologies

These conditions occur when there is an issue with the septation, positioning, or development of the great vessels of the heart. Most concerning is transposition of the great arteries (TGA) where the aorta is connected to the right ventricle and the pulmonary artery to the left ventricle. If this occurs in isolation, then the baby is cyanotic at birth. However, if it exists in conjuction with another defect (e.g. a septal defect) then admixture of the blood can occur, allowing some oxygenated blood to reach the systemic circulation (although the baby may remain cyanotic). Eventually, all patients require surgery to correct the vessels. In about 20% of infants, an *atrial septostomy* is performed as a bridging measure pending further surgery. This involves passing a balloon catheter through the venous system and through the foramen ovale; the balloon is inflated in the left atrium and pulled through the atrial septum to dilate the foramen ovale and tear the septum, ensuring that it does not close.

The other group of conditions are *outflow obstructions*. In the well child, these could be mild coarctation of the aorta (CoA), aortic stenosis (AS), or pulmonary stenosis (PS). In the unwell infant, one should consider severe CoA, hypoplastic left heart syndrome (HLHS), and aortic arch interruptions.

CoA accounts for about 5% of congenital heart defects (about 3/10,000 live births) and involves narrowing of the lumen of the aorta. It is classified according to the location of this narrowing relative to the ductus arteriosus: pre-ductal, ductal, and post-ductal. In severe pre-ductal constrictions, the circulation is duct-dependent as no blood can pass through the aorta prior to the duct and the circulation is dependent on blood bypassing the obstruction through the ductus arteriosus. This condition is usually managed with surgery.

Interruption of the aortic arch occurs when the proximal and distal segments of the aortic arch are not connected (this happens in about 1/10,000 live births). This is most likely due to malformations with the pharyngeal aortic arches – particularly the fourth – that contribute to the transverse segment of the arch of the aorta. The condition is classified according to the location of interruption relative to the three arteries that branch off the arch: distal to the left subclavian (type A), distal to the left common carotid (type B), and distal to the brachiocephalic (type C).

Mild-to-moderate stenosis of the aortic or pulmonary vessels will typically be asymptomatic and recognisable by a murmur on examination (ejection systolic murmurs). When severe, they may lead to cardiomegaly and heart failure; their treatment is with valve replacement.

▮ Tetralogy of Fallot

This condition presents with cyanosis either at birth or shortly thereafter. It is a right-to-left shunt that is frequently assessed in examinations. It is the most common cyanotic congenital heart malformation (occurring in 1/1000 live births). It has four key features: an overriding aorta (that receives blood from both ventricles), a large VSD, PS (causing right outflow obstruction), and right ventricular hypertrophy (as result of the stenosis). Babies that do not present with cyanosis in the first few weeks of life may be diagnosed in infancy with hypercyanotic spells and squatting on exercise. These spells are sudden episodes of profound cyanosis (occasionally with shock) due to hypoxia, and they typically occur during crying, defecating, or playing. The infant may squat as a compensatory mechanism to increase the peripheral vascular resistance in order to decrease the right-to-left shunt (by increasing the pressure in the left side of the system and reducing the pressure gradient).

Management of tetralogy of fallot (ToF) depends on the urgency of the presentation. The definitive intervention is surgery at about 6 months of age to correct the defects. If the baby is cyanotic in the neonatal period then an interim procedure is needed; either the right ventricular outflow tract can be dilated or a shunt can be placed between the subclavian artery and the pulmonary artery, known as a *Blalock–Taussig shunt*. The hypercyanotic spells are normally self-limiting but require medical management to correct the shock and metabolic consequences.

ToF can occur as part of *Alagille syndrome* where a mutation in the *Jagged1* ligand leads to cardiovascular and hepatic malformations.

▮ Patent Ductus Arteriosus

Approximately 10% of congenital heart disease is due to a patent ductus arteriosus (PDA). It is important to close the defect except in duct-dependent circulations as previously described. The reasons for this are twofold. First, persistent shunting of blood from the aorta to the pulmonary artery can lead to pulmonary hypertension. Second, there is a continuous increase in work for the left side of the heart to pump sufficient blood through the aorta to compensate for losses through PDA. As such, left untreated, one-third of patients die of heart complications by age 40 years, and two-thirds by age 60 years.

If recognised early by signs of tachycardia, shortness of breath, or a continuous 'machine-like murmur', then PGE1 levels can be reduced through use of non-steroidal anti-inflammatory drugs (COX inhibitors) to speed up the duct-closing process. Otherwise, patients require an operation to close the duct.

Blood Mixing

This is the result of arterial and venous blood mixing prior to expulsion via the outflow tract. It occurs in tricuspid atresia and atrioventricular septal defects (AVSDs). The newborn is cyanotic with shortness of breath. AVSD occurs in up to 10% of patients with Down's syndrome. Its management, as with septal defects, is dependent on its severity and is focused on interim measures prior to definitive surgical intervention at 3–6 months of age.

Tricuspid atresia is an absence of the tricuspid valve leading to a severely hypoplastic or absent right ventricle. The exact pathophysiology is unknown; however, blood flow through the linear tube is known to regulate the development and folding of the heart, and this deficit would severely affect this process. For the embryo to remain viable, tricuspid atresia needs to exist with an ASD (to transfer blood to the left atrium) and a VSD (to allow the left ventricle to pump blood back into the right side of the heart).

Procedures are performed to increase flow to the pulmonary circulation while reducing mixture. These include: the Blalock–Taussig shunt (discussed earlier); connection of the SVC to the pulmonary artery (*hemi-Fontan procedure* performed at age 6 months); and, later, connection of the IVC to the pulmonary artery (*Fontan procedure* performed at age 3–5 years).

Ebstein's Anomaly

In this condition, the tricuspid valves are positioned too inferiorly (towards the apex) and located within the upper aspects of the right ventricle, despite the annulus of the valve being in a normal position. It is very rare, with an incidence of about 1/200,000 live births. It presents over a very broad spectrum from infanthood to well into adulthood; this is dependent on its severity/location and the presence of other defects – commonly an ASD. In the presence of a severe malformation with ASD, patients present earlier with symptoms of right-to-left shunting (e.g. cyanosis). Where the defect is milder, patients present much later in life (sometimes aged in their 50s) with symptoms of right-sided heart failure.

Patent Foramen Ovale

Up to 25% of the population have a patent foramen ovale and the vast majority are asymptomatic. However, due to the communication between the pulmonary circulation and systemic circulation, it may then present with a *paradoxical embolus*.

This occurs when a venous thrombus embolises and passes through the foramen to deposit within the arterial circulation. This causes either a stroke or ischaemia. If the causative clot was analysed histologically, one would find that it would appear under the microscope to be different in aetiology from an arterial thrombus, with venous clots having more numerous red blood cells and fewer platelets.

RELEVANT MOLECULES

- *NKX2.5, GATA4,* and *TBX5:* genes required for development of the heart
- *Notch1:* this receptor regulates the epithelial-to-mesenchymal transition of cells in the endocardial cushions
- *Jagged1:* the ligand for the *Notch1* receptor
- *Retinoic acid:* promotes growth of endocardial cushions from the dorsal and ventral walls of the heart to divide the chambers into atrial and ventricular segments
- *PGE1:* serum levels of placental PGE1 drop at birth, and this leads to closure of the ductus arteriosus

KEY POINTS

- The heart is the earliest functional organ to develop and begins to contract at day 22.
- The endocardium and myocardium are formed from mesodermal tissue.
- Neural crest cells contribute to the formation of the aortic and pulmonary valves.
- Folding involves reorganising the linear heart tube to correctly position each segment.
- The foramen ovale forms due to apoptosis of the superior wall of the septum primum.
- Right-to-left shunting is essential in the fetus for ensuring oxygenated blood reaches the systemic circulation; it is reversed at birth leading to closure of the foramen ovale.
- The arch of the aorta receives contributions from the heart tube, pharyngeal aortic arches, and embryonic dorsal aortae.
- The ductus arteriosus shunts oxygenated blood from the pulmonary artery to the aorta.
- The ductus venosus shunts oxygenated and nutrient-rich blood from the umbilical vein directly into the IVC to bypass the liver.
- It is important to recognise congenital cardiac malformations that are dependent on shunting through the ductus arteriosus in order to prevent its closure.
- Severe right-to-left shunt defects present with cyanosis immediately following birth.
- Septal defects are the most common congenital malformation.

10 LIMB DEVELOPMENT

MESODERM AND ECTODERM

The production of the limbs is deceivingly complex as it requires a high degree of patterning to establish position, axes, and function. From the fourth to the eighth week of development, growth and patterning of the limbs occurs. It is initiated in response to *HOX* genes that signal the correct position of upper and lower limbs. These, in turn, lead to expression of *TBX5* genes to initiate growth in the upper limb and *TBX4* for the lower limb. This signals mesodermal and ectodermal tissue to organise and grow in response.

The mesoderm provides tissue through its paraxial and lateral plate parts. Recall that the somites are formed from paraxial mesoderm and divide into a sclerotome and dermomyotome. The dermomyotome divides into a *dermatome* that forms the dermis and a *myotome* that generates the muscles of the limb. The lateral plate mesoderm provides the bones, blood vessels, and connective tissue. Meanwhile, the ectoderm will provide the epidermis, nails, hair, and associated nervous system.

LIMB PROXIMAL-DISTAL OUTGROWTH AND PATTERNING

Initially, the limbs are small buds of mesenchymal tissue surrounded by ectoderm on their outer surface. They grow outwards in response to fibroblast growth factor (FGF) signals from the *apical ectodermal ridge* (*AER*), a small area of ectodermal cells at the tip of the bud (Figure 10.1). The AER uses FGF signals to encourage proximal-to-distal growth of the limb bud, with HOX genes specifying the distinct elements of the limb (e.g. HOX11 for the forearm and HOX12 for the carpal bones). The more proximal sections of the limb will differentiate first, as the cells in closest proximity to the AER and FGF remain undifferentiated, creating a *progress zone*.

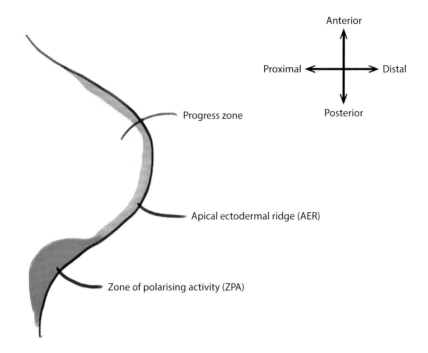

Figure 10.1 The limb bud growing in three axes in response to signals from the apical ectodermal ridge (distal) and the zone of polarising activity (posterior). The progress zone reflects the area where cells are becoming differentiated.

DORSAL-VENTRAL PATTERNING

The AER demarcates the boundary between dorsal and ventral aspects of the limb. Its position in the midline occurs as a result of a difference in patterning signals. Dorsally, the presence of *Wnt7* ensures that extensor structures form there, while bone morphogenetic protein (Bmp) promotes the growth of flexor side muscles and structures. These two morphogens form a gradient by antagonising one another.

CRANIO-CAUDAL PATTERNING

Patterning in this axis ensures that that the bones of the limbs are in the appropriate order, with caudal structures (e.g. the little finger) developing before cranial structures (e.g. the thumb). The region responsible for this is known as the *zone of polarising activity* (ZPA) and is located posteriorly/caudally. The morphogens being secreted from this area are Sonic hedgehog (Shh) and retinoic acid; they provide positional information for the cells in the limb by creating a diffusion gradient.

The molecules are dependent on the dorso-ventral morphogens. The gradient is supported by *Fgf8* secreted by the AER, which upregulates Shh. This in turn stimulates *Fgf4* in the caudal aspects of the limb.

DIGITALISATION

Regulated and direct apoptosis of cells in the distal limb bud results in the formation of digits. This process is dependent on Bmp signals interrupting the Shh signals from the ZPA.

CLINICAL SIGNIFICANCE

Anomalies in limb formation can occur due to either disruption of growth or errors in patterning. These processes are heavily dependent on the correct expression of genes and appropriate signalling of morphogens.

◼ Limb Bud Outgrowth Deformities

The limb bud grows out in response to *FGF* signalling from the AER, and disrupting this gradient leads to deformities dependent on the timing of the signalling loss. If it occurs early, it results in *amelia* and loss of the entire limb. Lower FGF concentrations or later loss of signal can result in either *meromelia* or *phocomelia*; the former is the absence of a segment of the limb (e.g. forearm), and the latter is the shortening of the limb.

◼ Digit Deformities

If *Fgf* signalling is lost following the majority of bud outgrowth, then it results in *adactyly* (loss of digits) or *ectrodactyly* (absent middle finger).

If there is a disruption in the signalling of Shh from the ZPA or surrounding *Bmp* levels then it results in *syndactyly* (fusion of digits); this most commonly affects the third, fourth, and fifth digits. When there is upregulation of the Shh levels then it results in *polydactyly* (an increased number of fingers).

◼ Thalidomide

During the late 1950s and early 1960s, thalidomide was used as an anti-emetic for pregnancy-associated 'morning sickness', as well as an anxiolytic in other patients (who could have been pregnant). It was found to be a *teratogen*: a molecule or substance that causes malformation of an embryo. Its use disrupted the progress zone and Fgf signalling from the AER, which resulted in phocomelia. The issues associated with this drug led to international reforms in drug regulation and marketing.

◼ Achondroplasia

This condition is also known as *dwarfism*. It results from an autosomal dominant mutation in the *FGF3* receptor gene that leads to a reduced response to signals from the AER, resulting in shortened limbs.

RELEVANT MOLECULES

- *HOX:* family of genes responsible for patterning the embryo
- *TBX5:* gene required for initiating upper limb development
- *TBX4:* gene required for initiating lower limb development
- *FGF:* releases signals from the apical ectodermal ridge to promote outward growth of the limb bud
- *Fgf8:* morphogen secreted by the AER along a dorso-ventral axis to upregulate Shh and position cranial portions of limb
- *Fgf4:* stimulated by Fgf8 to help differentiate caudal aspects of limb
- *Wnt7:* responsible for signals promoted dorsally in the limb bud to promote development of extensor structures
- *Bmp:* responsible for ventral signals to promote flexor structure development; also involved in digitalisation of the limb bud by disrupting Shh and promoting cell apoptosis
- *Shh* and *retinoic acid:* morphogens secreted by the ZPA (in the posterior/caudal limb bud), which is responsible for cranio-caudal patterning

KEY POINTS

- The limb is a highly patterned structure of the body whose development is dependent on positional signals.
- The expression of morphogens leads to the formation of gradients in the axis to signal to limb bud cells for differentiation that is correct in orientation and size.
- Disruption of morphogen signalling can lead to either the shortening or absence of structures in the newborn.

11 BONE DEVELOPMENT

Bone development, or osteogenesis, is a fundamental process that forms the skeletal system, essential for supporting the body, protecting vital organs, and enabling movement. This process involves two primary mechanisms: intramembranous ossification and endochondral ossification, each contributing to the formation of different bone types (Figure 11.1).

INTRAMEMBRANOUS OSSIFICATION

Intramembranous ossification is responsible for forming the flat bones of the skull, mandible, and clavicle. Unlike endochondral ossification, which relies on a cartilage model, intramembranous ossification occurs directly within *mesenchymal tissue*, a

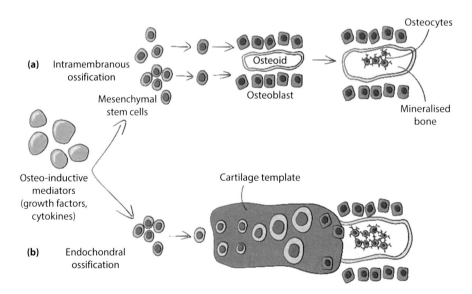

Figure 11.1 Diagram of (a) intramembranous ossification and (b) endochondral ossification.

100 EMBRYOLOGY EXPLAINED DOI: 10.1201/9781003479529-11

type of loose connective tissue rich in stem cells. This direct path to bone formation is essential not only for initial skeletal development but also for the lifelong process of bone repair and remodelling.

Mesenchymal stem cells (MSCs) within the embryonic fibrous connective tissue converge and cluster in regions predetermined to become bones. The exact origin of these cells is under debate, but they are mesodermal. These cells, under the influence of specific genetic and molecular cues, notably runt-related transcription factor 2 (RUNX2) and SRY-box transcription factor 9 (SOX9), then differentiate into *osteoblasts* – the bone-building cells. Osteoblasts secrete *osteoids*, an organic bone matrix composed primarily of type I collagen, which forms the initial scaffold for bone formation.

As the osteoid undergoes mineralisation, it traps some osteoblasts within it; these encased cells mature into *osteocytes*, which play a crucial role in maintaining bone tissue. The mineralisation process involves the deposition of calcium and phosphate ions into the osteoid, transforming it into a hard, durable bone matrix. This marks the creation of *woven bone*, a temporary and mechanically weaker form of bone that will eventually be replaced by mature lamellar bone through the bone remodelling process. Simultaneously, the developing blood vessels infiltrate the forming bone, promoting further ossification and the development of the bone marrow cavity. This vascularisation is crucial for the survival of bone tissue and serves as a conduit for the delivery of nutrients and removal of waste products.

The importance of intramembranous ossification is not limited to the embryonic phase but continues into adulthood, significantly influencing the repair mechanism of bone fractures. This ossification process is integral to the body's ability to regenerate and heal broken bones, effectively replicating the early bone formation stages to mend the skeletal structure. Specifically, in the event of a fracture, intramembranous ossification is activated at the site of injury, facilitating the direct deposition of bone matrix by osteoblasts derived from MSCs in the periosteum.

ENDOCHONDRAL OSSIFICATION

Endochondral ossification is the pathway through which the majority of the bones in the human body, such as the long bones (e.g. the femur and humerus), are formed. This process intricately replaces cartilage with bone in the developing embryo. It starts with the formation of a *cartilage model*, which outlines the future bone's shape. This model is made from *chondrocytes*, cells that produce a rich extracellular matrix composed of collagen and other substances, establishing a scaffold for future ossification. In the early stages of embryonic development, these chondrocytes proliferate rapidly within the *mesenchyme* (recall that this is the loosely organised, mostly mesodermal embryonic tissue from which various tissues can differentiate).

These chondrocytes also originate from MSCs, which aggregate then differentiate under the direction of transforming growth factors beta (Tgfs-β) and fibroblast growth factors (FGFs). *Chondroprogenitor cells* differentiate into mature chondrocytes, which actively produce and secrete the components of the cartilage extracellular matrix, including collagen (primarily type II collagen) and proteoglycans. This matrix is essential for cartilage's structural integrity and function. As chondrocytes continue to secrete the extracellular matrix, they become embedded within it, creating the cartilaginous structure. In the context of endochondral ossification, this cartilage model will eventually be replaced by bone tissue, but the initial formation of the model is crucial for determining the future bone's size and shape.

As endochondral ossification progresses, the centre of the cartilage model undergoes significant changes; chondrocytes enlarge and signal the surrounding matrix to calcify, subsequently dying due to the lack of nutrients passing through the calcified matrix. This calcification creates cavities within the cartilage, now primed for invasion by blood vessels from the surrounding perichondrium, transforming it into a vascularised periosteum, the outer fibrous layer of the bone. Blood vessels bring osteoblasts into the heart of the cartilage model, where these bone-forming cells begin to lay down bone matrix, creating the *primary ossification centre*. Osteoblasts generate bone on the remnants of the calcified cartilage matrix, initiating the formation of *trabecular bone*. Meanwhile, *osteoclasts*, large cells that can break down bone tissue, remodel the newly formed bone and mediate the expansion of the medullary (bone marrow) cavity.

Growth continues as chondrocytes at the ends of the bone – the *epiphyseal plates* – continue to proliferate, lengthening the bone. Ossification follows this cartilage proliferation, moving the zone of ossification further from the centre of the bone, allowing for continued bone lengthening until early adulthood. Finally, when the bone reaches its full size, the chondrocytes in the epiphyseal plate cease dividing, and the epiphyseal plate ossifies – a process known as epiphyseal closure – marking the end of bone growth in length.

These cells are regulated by a fine balance of many signals and proteins. Wnt proteins are crucial regulators of bone formation, primarily through the canonical Wnt/β-catenin signalling pathway, which promotes the differentiation and proliferation of osteoblasts. This signalling enhances the production of bone matrix proteins and supports the commitment of mesenchymal stem cells to the osteoblast lineage. Additionally, Wnt proteins inhibit *osteoclastogenesis* by inducing the production of *osteoprotegerin* (*OPG*), a decoy receptor that binds to and neutralises *receptor activator of nuclear factor κB ligand* (*RANKL*), preventing it from stimulating the maturation of bone-resorbing osteoclasts. RANKL is a critical molecule for osteoclast differentiation and activation, and its regulation by Wnt signalling ensures a balance between bone formation and resorption. Wnt signalling is also vital for bone repair and regeneration, responding to injury by facilitating the recruitment and differentiation of osteoprogenitor cells. Overall, the regulation of Wnt signalling, including its modulation of RANKL activity, is essential for maintaining bone mass and ensuring healthy bone development and homeostasis.

Osteogenesis is regulated by a host of signalling molecules and pathways, including bone morphogenetic proteins (BMPs), FGFs, parathyroid hormone-related protein (PTHrP), and Indian hedgehog (Ihh), each playing a pivotal role in coordinating the proliferation, differentiation, and activity of chondrocytes and osteoblasts throughout bone development.

CLINICAL SIGNIFICANCE

▌ Congenital Bone Disorders

Genetic mutations affecting the pathways and factors involved in bone development can lead to congenital bone disorders. Conditions such as *osteogenesis imperfecta*, characterised by brittle bones due to collagen defects, and *achondroplasia*, the most common form of dwarfism resulting from *fibroblast growth factor receptor 3 (FGFR3)* mutations, underscore the significance of genetic regulation in skeletal health. Early diagnosis through genetic screening and an understanding of the underlying molecular mechanisms are key to managing these conditions.

Achondroplasia

This the most common form of short-limbed dwarfism, caused by a genetic mutation in the FGFR3 gene, which leads to an abnormality in FGFR3. This receptor, when mutated, inhibits chondrocyte proliferation in the growth plates, significantly impacting endochondral ossification and resulting in shortened bones, particularly in the limbs and spine. As a result, there is limited lengthening at the growth plates. Initially, it was believed this process did not affect intramembranous ossification, leading to the relatively large head (macrocephaly) but absolutely regular-sized skull in patients with achondroplasia. However, recent research into FGFR3 gain-of-function mutations suggests that the macrocephaly may be a result of an additional impact on intramembranous ossification causing frontal bossing (enlargement of frontal bone) and macrocephaly of the skull. The clinical management of achondroplasia focuses on monitoring and addressing complications, such as spinal stenosis and hydrocephalus, and supporting the individual's physical and social development.

Collagen defects

Notably seen in conditions such as osteogenesis imperfecta, collagen defects highlight the importance of the organic matrix in bone strength and integrity. Collagen, particularly type I, is a critical component of the bone matrix, providing tensile strength and a framework for mineralisation. Mutations in the genes responsible for collagen production can lead to brittle bones that fracture easily, a hallmark of osteogenesis imperfecta. Furthermore, another common characteristic of these patients is blue discolouration of the white of the eyes (sclera) secondary to the thin scleral collagen, permitting the dark choroid vasculature to be seen. Management strategies focus on preventing fractures, promoting mobility, and improving bone density through pharmacological and physical therapies.

▌ Vitamin D deficiency

Deficiency in vitamin D can lead to *rickets* in children and *osteomalacia* in adults, conditions characterised by the softening and weakening of bones due to impaired bone mineralisation. Vitamin D plays a crucial role in calcium and phosphate metabolism, essential for healthy bone formation. The deficiency in this vitamin can be due to inadequate dietary intake, insufficient sunlight exposure, or problems with vitamin D metabolism. Treatment involves vitamin D and calcium supplementation, along with addressing the underlying causes to prevent future bone health issues.

RELEVANT MOLECULES

- *RUNX2:* a master regulator of osteoblast differentiation, essential for initiating bone formation during both intramembranous and endochondral ossification
- *SRY-box transcription factor 9 (Sox9):* plays a critical role in chondrocyte differentiation and cartilage formation, acting upstream of RUNX2 in the hierarchy of bone development
- *BMPs:* part of the Tgf-β superfamily, and key signalling molecules that promote the differentiation of MSCs into bone and cartilage
- *FGFs:* involved in various stages of bone development, including limb bud formation, growth plate chondrogenesis, and osteoblast differentiation
- *FGFR3:* gene for the Fgf3 receptor protein, in which a mutation leads to dwarfism
- *Wnt:* signalling proteins that regulate osteoblast proliferation and differentiation, crucial for bone mass accrual and maintenance
- *PTHrP:* maintains the growth plate chondrocytes in a proliferative state, delaying their hypertrophic differentiation in endochondral ossification
- *Ihh:* produced by prehypertrophic chondrocytes, this host signalling molecule interacts with PTHrP to regulate chondrocytes in osteogenesis
- *Type I collagen:* main structural protein in the extracellular matrix of bone, providing tensile strength and framework for mineralisation
- *OPG* and *RANKL:* regulate osteoclast differentiation and activity, critical for bone remodelling and calcium homeostasis
- *Vitamin D and its receptors:* essential for calcium absorption and bone health, its signalling influences osteoblast and osteoclast function

KEY POINTS

- Intramembranous ossification forms flat bones directly from mesenchymal tissue without a cartilage stage.
- Endochondral ossification creates long bones from a cartilage model that is eventually replaced by bone.
- Bone growth and remodelling continue throughout life, balanced by the actions of osteoblasts (forming bone) and osteoclasts (resorbing bone).
- Skeletal system formation begins with mesodermal MSCs differentiating into either osteoblasts or chondrocytes, depending on the ossification process.

12 FOREGUT: LUNG AND DIAPHRAGM

After folding and formation of the 'tube within a tube' structure, the endoderm becomes the innermost gut tube and runs from the oropharyngeal membrane (presumptive mouth) to the *cloacal membrane* (presumptive anus and urethral opening). This tube is divided into three parts with distinct blood supplies and derivates. It was initially believed that the arteries defined the segments of the gut tube; however, it is now known that the *HOX* genes code for the location of the arterial branches and organs.

The derivates and blood supplies of the gut tube are very commonly assessed in written examinations and anatomy spotters, so they have been listed in Table 12.1. For the foregut, it is important to remember that any organ or system which communicates with the gut tube is formed by the endoderm (e.g. the liver and pancreas), whereas organs that do not connect to the gut are not thus formed (e.g. the heart and spleen). The arteries are unpaired central branches of the descending aorta. These arteries will divide to supply all of the derivates of the relevant segment of the gut through a named artery.

Initially, the endodermal gut tube is surrounded by a layer of mesoderm, but this disperses to the edges to remain as the *dorsal mesentery*. This structure suspends the gut tube from the abdominal cavity walls within a protective casing known as the *peritoneum*. The peritoneum has two parts: the *visceral peritoneum* from the *splanchnic mesoderm* that covers the organs, and a *parietal peritoneum* from the *somatic mesoderm* that lines the abdominal cavity (Figure 12.1). What should also be evident here is that the dorsal mesentery is a double layer/fold of visceral peritoneum.

Accordingly, the position of organs relative to the peritoneum in development, and as final structures, is important embryologically and clinically. Some organs develop and exist inside of the peritoneum, and are known as *intraperitoneal*. Others develop and exist outside the peritoneum; these are known as *primary retroperitoneal* organs. The final group develop within the peritoneum to then lie outside of it; these are *secondary retroperitoneal* organs. This is another commonly

Table 12.1 The derivates and blood supplies of the gut tube

Subdivision	Derivates	Segment	Artery
Foregut	• Pharynx • Respiratory tract • Lungs • Larynx • Stomach • Duodenum (proximal two parts) • Liver • Pancreas	From the mouth to the second part of the duodenum (entry of the common bile duct)	Coeliac
Midgut	• Duodenum (distal two parts) • Ileum • Jejunum • Appendix • Caecum • Ascending colon • Transverse colon (proximal two-thirds)	From the entry of the common bile duct in the duodenum to two-thirds of the way along the transverse colon	Superior mesenteric
Hindgut	• Tranverse colon (distal third) • Descending colon • Sigmoid colon • Rectum • Anal canal (proximal part)	From two-thirds of the way across the transverse colon to the pectinate line of the anal canal	Inferior mesenteric

assessed set of facts in examinations. The organs that are retroperitoneal are listed below:

- *Primary*: kidney, adrenals, bladder, aorta, inferior vena cava, ureter, abdominal part of oesophagus, lower rectum, anal canal
- *Secondary*: ascending colon, descending colon, second part of duodenum, pancreas (head, body, neck), upper rectum

TRACHEA AND BRONCHIAL TREE

The trachea and lungs begin to develop from the endodermal gut tube and surrounding mesoderm at the fourth week of gestation. An endodermal respiratory diverticulum (bud) grows ventrally from the foregut. This is surrounded by

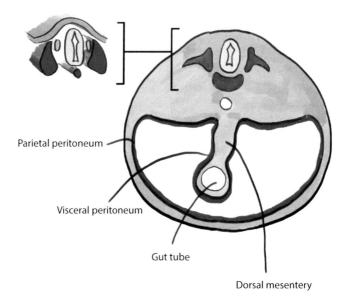

Figure 12.1 The dorsal mesentery of the mesoderm with a suspended endoderm gut tube.

tracheo-oesophageal ridges of mesoderm that separate the budding lung from the oesophagus. This process is dependent on retinoic acid signalling.

The endoderm begins to proliferate to form a rod of tracheal epithelium that recanalises (around the tenth week of gestation). The U-shaped cartilage, posterior smooth muscle, and connective tissue of the trachea are formed from the mesoderm. During the fourth week, the rod of tracheal epithelium will divide in two to form the *primary bronchi*, and then continue to bifurcate (Figure 12.2). The bifurcation of the endodermal buds occurs into the surrounding mesodermal tissue and is also regulated by retinoic acid levels.

LUNG DEVELOPMENT

Once the bronchial tree has been established, by the fifth week of gestation, the lung begins to form in four distinct stages. The first period is known as the *pseudoglandular* phase (weeks 5–16), in which the lung is a series of branched terminal bronchioles with no gas-exchange surfaces or alveoli. Therefore, the birth of a fetus at this phase is non-viable as respiration is not possible.

The *canalicular* period (weeks 16–26) follows and involves the formation of respiratory bronchioles and alveolar ducts with the emergence of a few terminal

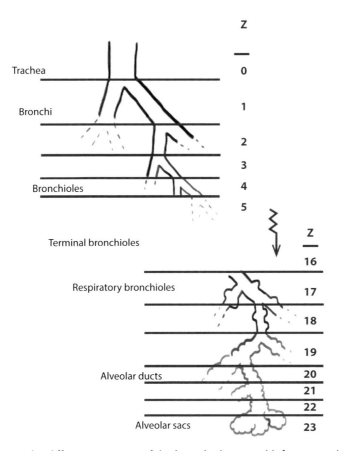

Figure 12.2 The different segments of the bronchial tree and bifurcation phases. 'Z' marks the number of branches in the logarithmic 2^Z.

sacs; these sacs resemble early alveoli with vascularisation for the exchange of gases. With the occurrence of sacs, *type II pneumocytes* develop around week 24 to release *surfactant*. The role of the phospholipid surfactant is to prevent the collapse of airways and alveoli. It does so by maintaining the surface tension of the airways and alveoli; otherwise gradients of pressure would form between small and large alveoli, leading to the collapse of the smaller alveoli. At this stage of development, respiration is poor but possible, and premature infants may survive, dependent on the presence of surfactant.

At this point, the basic outline of a respiratory tree is formed, and the remainder of lung development is for the maturation of the alveolar sacs into buds of alveoli: this is dubbed the *terminal sac* period (from week 26 to birth). There is a rapid increase in the number of terminal sacs/alveoli accompanied by thinning and increased vascularity of the *alveolar* walls to maximise potential gas exchange. This generation of alveoli and remodelling continues until 8 years of age, through a period of lung maturation known as the alveolar period; the child will ultimately have more than six times as many alveoli as it is born with.

As the lungs develop, they are separated from the heart during the fifth week. *Pleuropericardial folds* develop from the lateral body wall anterior to the developing lung. These folds grow to become *pleuropericardial membranes* that fuse in the midline to separate the pleural cavity (dorsal) from the pericardial cavity (ventral). These membranes become the fibrous pericardium.

DEVELOPMENT OF THE DIAPHRAGM

The diaphragm isolates the thoracic cavity from the abdominal cavity. It forms from many parts: the septum transversum, pleuroperitoneal membranes, mesentery of the oesophagus, and myoblasts from the third to fifth cervical somites. The septum transversum is initially the most cranial structure in the embryonic disc; however, cranio-caudal folding brings it to lie between the presumptive heart and liver. This positions it at the level of C1, and it then it grows caudally with the development of the lungs, taking myoblasts from the third to fifth somites with it. In doing so, it also takes the motor innervation of these myoblasts from C3, C4, and C5 to form the skeletal muscle component of the diaphragm. This is why the diaphragm is innervated by the phrenic nerve despite being a low thoracic structure ('C3, 4, and 5 keep the diaphragm alive').

When this completes, there remains a channel between the pericardial area and the peritoneum, known as the *pericardioperitoneal canal*. Two *pleuroperitoneal folds* begin to grow from the lateral walls of the body to close this canal. They meet with the oesophageal mesentery and septum transversum to form *pleuroperitoneal membranes*.

CLINICAL SIGNIFICANCE

▌ Watershed Areas of the Gut

Watershed areas refer to zones that receive a dual blood supply from the distal end-branches of two large arteries. In the gut, this typically refers to the *splenic flexure* (*Griffiths point*) that marks the boundary between the midgut and the hindgut. It receives blood supply from the end arteries of the superior and inferior mesenteric branches of the aorta. It can also refer to the *rectosigmoid junction* (*Sudeck's point*) between the zones of the inferior mesenteric and superior rectal arteries. During systemic hypoperfusion (e.g. during haemorrhage or shock), these areas are susceptible to ischaemia.

▌ Retroperitoneal Organs

The peritoneum is known as the 'policeman of the abdomen' because it migrates towards areas of infection/inflammation and puncture sites in stab wound injuries. This aims to mitigate the injury by reducing the zone of inflammation and infection. In stab injuries, it covers the puncture site to reduce additional risk of infection from the outside. An understanding of the location of organs relative to the peritoneum is important for surgeons to identify structures intra-operatively. Furthermore, it has led to the development of direct procedures for retroperitoneal organs without affecting other organs or risking intraperitoneal infection; these include insertion of a suprapubic catheter, a renal biopsy, and an ascitic tap.

▌ Tracheo-Oesophageal Fistula

If the tracheo-oesophageal septum fails to separate the oesophagus from the trachea, then a fistula can occur. Typically, the oesophagus is interrupted to form a proximal and distal end, with the proximal remaining as a short blind-ended sac and the distal end adjoining the trachea via a fistula; however, many variations of this occur. The newborn will present with choking, coughing, vomiting, and cyanosis whenever they try to feed.

This condition can occur as part of a wider syndrome of associated conditions known as the *VACTERL* association. These patients will have at least three of the following features:

- **V**ertebral anomalies (spinal cord deformity)
- **A**nal atresia (imperforate anus)
- **C**ardiovascular abnormalities
- **T**racheo-oesophageal fistula
- (O)**E**sophageal atresia
- **R**enal abnormalities
- **L**imb anomalies

❚ Respiratory Distress Syndrome

This condition, also known as *hyaline membrane disease*, is due to an absence or inadequate production of surfactant. It is usually caused by a lack of type II pneumocytes, leading to insufficient surfactant production and collapse of the airways with a glossy hyaline membrane. Mothers of babies at risk of being born prematurely are given steroids to promote lung development and the formation of type II pneumocytes.

❚ Congenital Diaphragmatic Hernia

Failure of the pleuroperitoneal membranes to close the pericardioperitoneal canals can lead to a *congenital diaphragmatic hernia* (1/2000 live births), where abdominal viscera protrude into the thoracic cavity. In the majority of cases, the herniation is on the left aspect of the aortic or oesophageal hiatus (85%) since the liver obstructs the route on the right. This is also known as a *Bochdalek hernia*. The presence of the abdominal contents within the thoracic cavity can lead to pulmonary hypoplasia and difficulty in breathing. When the defect is ventral, then it is known as a *Morgagni hernia*.

RELEVANT MOLECULES

- *HOX:* expression of these genes demarcates the gut into the foregut, midgut, and hindgut
- *Retinoic acid:* signalling molecule which guides the separation of the trachea from the oesophagus and the development of the bronchial tree

KEY POINTS

- The gut is divided into three regions: foregut, midgut, and hindgut.
- Abdominal viscera are either intraperitoneal, primary retroperitoneal, or secondary retroperitoneal.
- The lung is predominantly an endodermal foregut structure.
- Phases of lung development reflect the presence of gas-exchange surfaces.
- Surfactant is essential to maintain surface tension and prevent the collapse of airways.
- The viability of the fetus is linked to the development of the lungs.

13 FOREGUT: OESOPHAGUS AND STOMACH

The gastrointestinal tract is suspended from the abdominal wall via a series of mesenteries. The dorsal mesentery connects to the length of the gut with the exception of the oesophagus and anus, which are directly in contact with the body wall. The liver and stomach have additional mesenteries which regress to form adult structures: the *ventral mesentery* of the liver will become the *falciform ligament*, the dorsal mesentery of the stomach will form the *greater omentum*, and the ventral mesentery between the stomach and liver will become the *lesser omentum*.

DEVELOPMENT OF THE OESOPHAGUS

The oesophagus develops as a solid rod of thoracic foregut at the end of the fourth week of gestation. This occurs after the respiratory diverticulum buds ventrally and the tracheo-oesophageal ridges separate the presumptive oesophagus from the presumptive trachea. It elongates rapidly relative to the rest of the gut tube – being extended by the rapidly growing pharynx – in order to create the distance needed to reach through the thoracic cavity. The structure canalises around week 9 to form a tube.

DEVELOPMENT OF THE STOMACH

The stomach forms as a result of rotation, differential growth, and positional change of the gut tube. The process starts with a spindle-shaped dilatation of the foregut distal to the septum transversum. The dorsal aspect begins to grow faster than the ventral side to create the greater curvature. Further expansion superior to the greater curvature forms the fundus and cardiac incisure, leading to the distinctive shape of the stomach (Figure 13.1). As this happens, the thinning of the ventral mesentery leads to a 90-degree rotation in the cranio-caudal axis which:

- Brings the dorsal border to the left to form the *greater curvature*
- Places the ventral border to the right to form the *lesser curvature*

- Repositions the left-sided vagus nerve bundle to become the *ventral vagus plexus*
- Repositions the right-sided vagus nerve bundle to become the *dorsal vagus plexus*
- Prepares the dorsal and ventral mesenteries (mesogastrium) of the stomach to become the greater and lesser omentum
  - This rotation of the mesentery creates a space behind the stomach called the *lesser sac*
- Fuses the second part of the duodenum to the posterior wall, making it a secondary retroperitoneal organ

During this process, two mesodermal processes are occurring. The first is that smooth muscle is recruited and proliferates in the distal aspect of the stomach to form the *pyloric sphincter*. Secondly, the spleen begins to form within the *dorsal mesogastrium* (Figure 13.2).

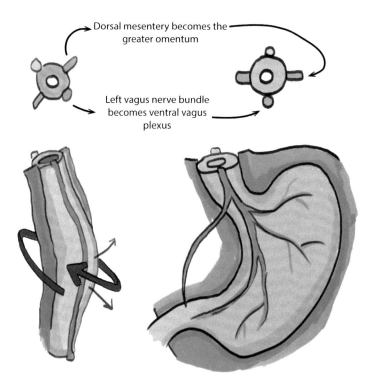

Figure 13.1 Expansion and rotation of the stomach.

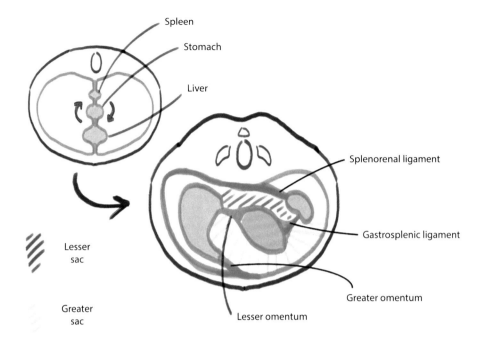

Figure 13.2 Formation of the greater and lesser sac with rotational movements of the stomach.

CLINICAL SIGNIFICANCE

▌ Polyhydramnios and Oesophageal Atresia

In utero, the fetus will use the surrounding amniotic fluid to aid the development of its internal organs. In order to prepare for breathing, it practises diaphragmatic movement to gain adequate strength and function to take a full breath at birth. This leads to an inflow and outflow of amniotic fluid through the lung tree, which stimulates the epithelium to promote maturation and development of the lung.

Polyhydramnios occurs in 1% of all births and occurs when there is an excess of amniotic fluid within the amniotic sac. It can be diagnosed during ultrasound scans of the fetus. It is a concern because it is associated with severe complications including cord prolapse, placental abruption, premature birth, and perinatal death. It typically occurs with either a systemic disease in the mother or a congenital condition/infection in the fetus, so the newborn should be carefully examined for any issues.

Following canalisation of the oesophagus, the fetus will swallow the amniotic fluid to aid development of the gastrointestinal tract. If canalisation of the oesophagus at week 9 fails then the fetus is unable to swallow intra-uterine amniotic fluid, leading to polyhydramnios. This is known as *oesophageal atresia* and is often associated with a *tracheo-oesophageal fistula*. When the blockage is not complete, then it is known as *oesophageal stenosis*. A similar condition can occur when the duodenum is not fully canalised, known as *duodenal atresia*. The newborn will regurgitate, cough, or choke on attempted feeding, due to an inability to swallow. As with any patient with an impaired swallow, this leads to a risk of aspiration pneumonia.

▌ Congenital Hiatal Hernia

In Chapter 12, we discussed congenital diaphragmatic hernias due to inadequate formation of the diaphragm. If the oesophagus does not adequately elongate through the thoracic cavity then the *cardia* of the stomach is pulled up into the thoracic cavity, leading to a hiatus hernia. The effects of this condition are similar to that of a congenital diaphragmatic hernia.

▌ Hypertrophic Pyloric Stenosis

In this condition, the smooth muscle at the outflow of the stomach (pylorus) is hypertrophic (i.e. overproliferated). This leads to an outflow obstruction on ingestion of food. The standard description is that the newborn will projectile vomit about 1 hour after feeding, and the vomitus is *non-bilous* (as it has not mixed with bile salts in the duodenum). On clinical examination, a small hard lump (often described as an *olive*) may be palpable inferior to the sternum. It occurs in about 0.5% of births, and requires a *pyloromyotomy* procedure. The danger to the newborn is inadequate hydration/nutrition.

RELEVANT MOLECULES

- None

KEY POINTS

- As the epithelium of the gut tube proliferates, it develops as a solid rod that requires canalisation; failure of this process leads to stenosis and atresia.
- The oesophagus forms due to elongation of the gut tube proximally, following separation from the presumptive trachea.
- The stomach forms as a result of rotation, differential growth, and positional change of the gut tube.
- The rotation of the stomach is very important in positioning many of the adult organs and is responsible for the creation of the intra-abdominal cavities.

14 FOREGUT: HEPATOBILIARY AND PANCREAS

The liver, gallbladder, and pancreas develop mainly from the endodermal foregut. The fetal liver acts as the first haematopoietic organ in the embryo until the third trimester, where the function is performed by the bone marrow.

DEVELOPMENT OF THE LIVER AND GALLBLADDER

The liver develops around day 22 as a small endodermal thickening overlying the ventral duodenum known as the *hepatic plate*. These endodermal cells will form the *hepatocytes*, *hepatic ducts*, *biliary cells*, and *bile canaliculi*; mesoderm from the septum transversum will generate the supporting stromal connective tissue, vascular sinusoids, and the *Kupffer cells*. This mesoderm is very important in inducing the formation of the hepatic plate as it releases bone morphogenetic protein (Bmp) to act on the nearby endoderm, which in turn makes this region highly sensitive to the fibroblast growth factor 2 (FGF2) signals from the cardiac mesoderm that induce hepatic development.

The hepatic plate will proliferate towards the septum transversum to form a *hepatic diverticulum* around day 26. Caudal to the hepatic duct, a *cystic diverticulum* develops that will go on to form the gallbladder and share a common bile duct with the hepatic diverticulum (Figure 14.1). It is important to note that, although it buds from the hepatic duct, the cystic diverticulum forms from a distinct group of cells on the ventral duodenum.

DEVELOPMENT OF THE PANCREAS

The pancreas forms from a *dorsal* and *ventral pancreatic bud*. The ventral bud grows as an outpouching from the hepatic duct, caudal to the cystic diverticulum (Figure 14.1). The dorsal bud proliferates from the dorsal duodenum, directly

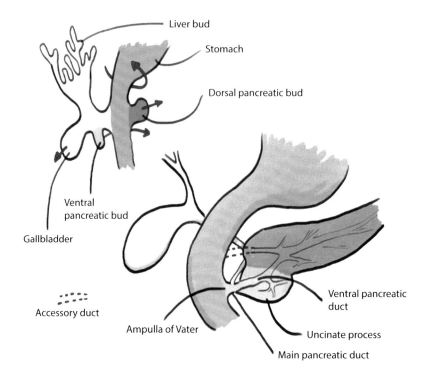

Figure 14.1 Development of the endodermal organs from the foregut.

opposite the site of growth of the hepatic diverticulum, and grows into the dorsal mesentery. The ventral bud will form the hook-like *uncinate process* of the pancreas; meanwhile, the dorsal bud will form the head, body, and tail of the pancreas.

As the stomach and midgut rotate, the repositioning brings together the dorsal bud with the ventral bud and common bile duct, such that they open through a common opening into the second part of the duodenum – the *ampulla of Vater*. The dorsal and ventral buds will then fuse at around the sixth week of development. The pancreatic duct within the dorsal bud usually degenerates, leaving the ventral duct as the *main pancreatic duct*; however, if there is inadequate fusion then accessory ducts may persist. The dorsal pancreatic duct often forms the *accessory pancreatic duct* (*duct of Santorini*), which can persist and may join the *main pancreatic duct* (*duct of Wirsung*).

The initiation of pancreatic development is dependent on the expression of *PDX1* genes, with expression of *PAX1* and *PAX6* specifically needed for differentiation of the endocrine (*islets of Langerhans*) cell lineages.

CLINICAL SIGNIFICANCE

▌ Biliary Atresia

Similar to other disorders leading to atresia, this occurs when there is proliferation of epithelial tissue lining the intra- and extra-hepatic biliary tree with failure of canalisation of the lumen. It occurs in about 1/14,000 live births and presents as one of the rarer causes of prolonged neonatal jaundice. Without intervention, it leads to progressive chronic liver failure that results in death. Once recognised, a *Kasai (hepatoportoenterostomy) procedure* can be performed within 60 days, in which a loop of jejunum is anastamosed to the porta hepatitis in order to allow biliary drainage. In about 80% of infants, this is successful if performed early enough, with the likelihood of success diminishing with age. Without this procedure, the infant requires a liver transplant; biliary atresia remains the most common indication for a hepatic transplant in the paediatric population.

▌ Annular Pancreas

This occurs when a ring of pancreatic tissue surrounds the second part of the duodenum and constricts the lumen. It results from a bifid ventral pancreatic bud

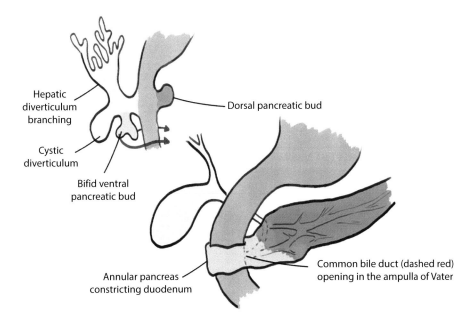

Figure 14.2 Bifid ventral pancreatic bud (iliac) segments each rotating in a different direction and fusing with the dorsal pancreatic bud (purple), leading to an annular pancreas.

or errors in the fusion process. In the case of the bifid bud, the tail of each ventral segment rotates in a different direction and adjoins the dorsal bud, leading to incomplete rotation and fusion (Figure 14.2). Its symptoms would be similar to that of hypertrophic pyloric stenosis with projectile vomiting occurring a short time (about 1 hour) after feeding. However, it is distinguished by the vomitus being bile-stained as the food will have mixed with the bile.

RELEVANT MOLECULES

- *Bmp:* signals released from the septum transversum mesoderm make the ventral duodenum responsive to Fgf2 signals
- *FGF2:* a signalling protein released by the cardiac mesoderm to induce hepatic development of the endodermal tissue
- *PDX1:* gene required for pancreatic development
- *PAX1* and *PAX6:* genes necessary for differentiation of pancreatic endocrine tissue

KEY POINTS

- The liver is endoderm-derived with the exception of the stromal connective tissue, vascular sinusoids, and Kupffer cells.
- The gallbladder forms from endoderm that is caudal and distinct from the hepatic tissue.
- The pancreas forms from a ventral bud and dorsal bud.
- The dorsal bud forms the head, neck, and tail, while the ventral bud forms the uncinate process of the pancreas.
- The main pancreatic duct is derived from the ventral bud, but it is common for accessory ducts to persist.

15 MIDGUT DEVELOPMENT

The midgut runs from the distal half of the duodenum to two-thirds of the way across the transverse colon. It starts precisely after the point of entry of the common bile duct at the ampulla of Vater. Its development is dependent on differential rates of growth and directional rotation.

GROWTH AND HERNIATION

Rapid growth of the midgut portion of the gastrointestinal tube leads to the formation of a *primary intestinal loop* that is shaped like a hairpin. This has cranial and caudal halves (limbs) along a central axis defined by the vitelline duct and superior mesenteric artery (Figure 15.1). The cranial half will form the distal duodenum, jejunum, and proximal ileum, while the caudal half forms the distal ileum, appendix, caecum, ascending colon, and the first two-thirds of the transverse colon.

The *vitelline duct* in the embryo is a communicating tract between the yolk sac and midgut. It receives a blood supply from the left and right vitelline arteries and acts as a transit of substances from the yolk sac into the embryo; however, the function in humans is not well understood. The left vitelline artery will involute, while the right will become part of the *superior mesenteric artery*. The duct is usually obliterated by the eighth week of gestation, but may persist as a *Meckel's diverticulum*.

Around the sixth week of gestation, the primary intestinal loop and liver are growing at a greater rate than the surrounding structures and expand to a volume beyond the size of the fetus's abdominal cavity. This leads to a herniation of the midgut through the presumptive umbilicus (where the vitelline duct runs), known as the *physiological umbilical herniation*.

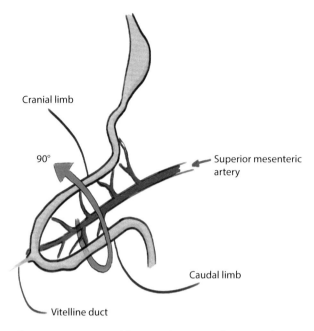

Figure 15.1 The primary intestinal loop and its axis of rotation/symmetry.

ROTATION AND RETRACTION

As the loop herniates, it undergoes a primary rotation of *90 degrees counterclockwise* around the axis of the superior mesenteric artery (Figure 15.2). This moves the cranial limb to the right and the caudal limb to the left. While doing so, it grows further and forms the jejunal loops and the vermiform (worm-like) appendix. Around week 11, the primary intestinal loop begins to retract. It is not fully understood if this is an active process due to regression and pulling, or a passive process as a result of the relative growth of the surrounding abdominal cavity.

As the bowel loop is retracted into the abdominal cavity, it undergoes a further rotation of *180 degrees counterclockwise* (a total of 270 degrees counterclockwise during herniation and retraction). This is important for the final positioning of the organs as it places the cranial limb to the left and the caudal limb to the right, resulting in locating the appendix in the right lower quadrant. The process is completed by the eleventh week.

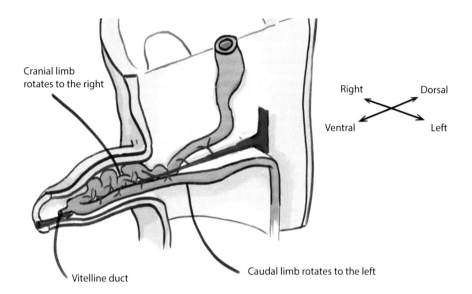

Figure 15.2 The herniated primary intestinal loop following growth of the presumptive small bowel.

CLINICAL SIGNIFICANCE

▌ Meckel's Diverticulum

Failure of the vitelline duct to be obliterated will result in a Meckel's diverticulum. This is of relative importance in clinical medicine as it is a differential diagnosis for abdominal pain, particularly that migrating to the right lower quadrant (similar to appendicitis). It can be found in 2% of individuals and is the most common congenital abnormality of the gut. The persistent vitelline duct can exist as an outpouching of the small intestine (Meckel's diverticulum), a cyst (with two closed ends), or a fistula (a connection between two epithelial-lined organs) (Figure 15.3).

A Meckel's diverticulum can be considered a true diverticulum as it contains all three layers of the gastrointestinal tract. It typically contains gastric tissue, and this is identifiable using a 99-technetium (nuclear radioisotope) scan.

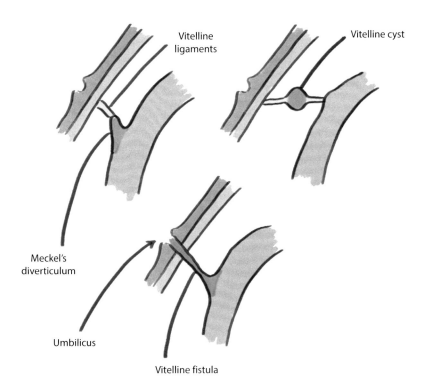

Figure 15.3 The differential forms of persistent vitelline duct tissue as a diverticulum, cyst, and fistula.

The aetiology and pathophysiology of a Meckel's diverticulum can be recalled by the 'rule of twos':

- 2% of the population
- 2 inches in length
- 2 feet proximal to the ileocaecal valve
- 2 years old is the most common age of presentation
- 2:1 male-to-female ratio of incidence
- 2 types of common ectopic tissue (gastric and pancreatic)

Omphalocele, Gastroschisis, and Umbilical Hernia

An *omphalocele*, also known as *exomphalos*, is a herniation of the bowel (or other viscera) through the umbilicus and covered by a thin avascular membrane. It is a rare abdominal wall defect occurring in about 1/4000 births and it is associated with a high mortality rate. It is thought that a potential mechanism is malrotation during retraction of the midgut into the abdominal cavity, which leads to insufficient closure of the ventral abdominal wall. Alternative causes have been described, including the failure of somitic myotomes to form abdominal muscles.

This is in contrast to *gastroschisis* in which there is herniation of the bowel at a site other than the umbilicus *without a covering sac*. It typically occurs to the right of the umbilicus and at a similar rate (about 1/2500 births) to omphaloceles. Similarly, there are many theorised causes but these focus on failures of adequate abdominal wall formation rather than gut rotation/retraction.

An *umbilical hernia* is a small, skin-covered protrusion of bowel or omentum through the umbilicus. It can occur congenitally due to an inadequate meeting and closure of the abdominal wall around the umbilicus. This is in contrast to the adult umbilical hernia that occurs due to an acquired weakness of the abdominal wall (e.g. obesity, post-operative, or post-partum). In the neonate, it is most obvious when the baby cries as this increases the intra-abdominal pressure and protrudes the viscera. It is typically repaired at the age of 5 years. All forms of hernias are repaired electively (or as an emergency) due to the potential of the bowel to become stuck (non-reducible) or rotate, leading to strangulation, ischaemia, and necrosis.

Intestinal Malrotations

These congenital anomalies occur in about 1/3000 live births and the presentation is related to the final position of organs (not the degree of malrotation). It can lead to midgut volvulus, in which the infant will present with symptoms of abdominal pain, sudden episodes of crying (due to cramps), constipation, or vomiting. The infant will undergo a *Ladd procedure* to remove the bands across the bowel. The bowel is then re-inserted into the abdomen with the small bowel on the right and large bowel on the left. The patient leads a normal life; however, if the infant develops appendicitis as an adult, it can lead to left-sided abdominal pain. Note that this is different to *Ladd's bands,* which are constricting bands of tissue that are a common cause of adult small bowel obstruction, typically caused by abdominal surgery or severe inflammatory pathology.

Similar to the post-Ladd procedure abdomen, a midgut which undergoes no rotation (*intestinal non-rotation*) will result in the small bowel being right-sided and the large bowel on the left. These patients usually have no symptoms. A small proportion develop signs as a result of kinking of the duodenum by the misplaced overlying bowel, which can lead to a blockage.

RELEVANT MOLECULES

- None

KEY POINTS

- The midgut runs from the distal half of the duodenum to two-thirds of the way across the transverse colon.
- Increased growth of the midgut leads to the formation of a primary intestinal loop.
- This loop grows faster than the surrounding abdominal cavity and so herniates through the umbilicus.
- The herniated loop undergoes 90 degrees of counterclockwise rotation.
- As it retracts, it undergoes a further 180 degrees of counterclockwise rotation.
- In total, the midgut will rotate 270 degrees in a counterclockwise direction.

16 HINDGUT AND BLADDER DEVELOPMENT

The hindgut will form the final third of the transverse colon, descending colon, sigmoid colon, rectum, and upper anal canal. It is supplied by the inferior mesenteric artery and its branches. The rotation and retraction of the midgut will position the hindgut appropriately. At its distal end, it terminates as an expanded pouch known as the *cloaca*, which will divide into two to contribute to the lower urogenital tract and the anorectal canal.

ENTERIC NERVOUS SYSTEM

The bowel is innervated by an *enteric nervous system* that forms as a result of the migration of neural crest cells. It is a very complex neural plexus that is divided into the *submucosal (Meissner's) plexus* and *myenteric (Auerbach's) plexus*. It can operate independently of any central nervous system stimulation so is often dubbed the 'second brain'.

The myenteric plexus lies between the circular and longitudinal muscle layers of the bowel wall and is responsible for the modulation of peristalsis by controlling the tone and frequency of contractions. It provides both sympathetic and parasympathetic innervation to the gut. The submucosal plexus is found in the submucosa (between the mucosa and muscularis propria) and innervates the gut mucosa. It is responsible for the regulation of bowel secretions, local absorption, and local contraction.

FORMATION OF THE ANORECTAL CANAL AND BLADDER

The hindgut ends in an endodermal-lined pouch known as the *cloaca* whose opening is covered by a cloacal membrane (the presumptive anus). Within this structure, the endoderm of the hindgut meets the external ectoderm, marked by the *anal pit*.

Lying ventrally is the allantois, which is connected to the yolk sac. Its function in the human is not well understood, but it is used by reptiles and birds (which do not have placentas) for respiration and the removal of waste products. It is hypothesised that in mammals it contains mesodermal tissue, which is required for the formation of the umbilical arteries (to carry deoxygenated blood to the placenta) and to remove nitrogenous waste from the bladder. In the human, its proximal portion (the *urachus*) will form the superior part of the bladder, and the remainder regresses into a dense fibrous cord (the *median umbilical ligament*) that connects the urinary bladder anteriorly to the umbilical region.

The expanded cloaca is divided by the growing urorectal septum into a *ventral urogenital sinus* and a *dorsal anorectal canal* (Figure 16.1). The septum starts as two lateral segments that fuse in the midline and grow caudally. They meet with the cloacal membrane around the seventh week of development, to separate it into a *urogenital membrane* ventrally and an *anal membrane* dorsally. The anal membrane will perforate in the ninth week to form the anus. The urorectal septum will remain in the newborn as the *perineal body*, separating the genitalia from the anus as the perineum.

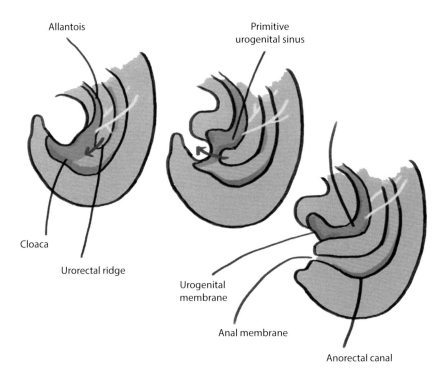

Figure 16.1 Division of the cloaca into the urogenital sinus and anorectal canal.

The endodermal anorectal canal will form the rectum and the upper two-thirds of the anal canal. This meets with the lower ectodermal anal canal at the *pectinate line* (also known as the *dentate line*). This is important, as it marks a distinction between the innervation, lymphatics, and vascular supply of the two regions, as outlined below:

- Upper anal canal
  - *Artery:* superior rectal artery of the inferior mesenteric branch of the aorta
  - *Vein:* superior rectal vein of the inferior mesenteric (draining into the *portal venous system*)
  - *Lymphatics:* internal iliac nodes
  - *Nerve:* visceral innervation via the *inferior hypogastric plexus*
  - *Sensation:* only to stretch
- Lower anal canal
  - *Artery: inferior rectal artery* of the internal pudendal branch of the internal iliac artery
  - *Vein: inferior rectal vein* draining into the systemic circulation via the internal pudendal vein
  - *Lymphatics:* superficial inguinal nodes
  - *Nerve:* somatic innervation via the inferior rectal branches of the *pudendal nerve*
  - *Sensation:* touch, pain, temperature, and pressure

A highly variable structure is the *middle rectal artery*. It is found in the majority of people and has been shown to supply both the upper and lower parts of the canal in different studies. It is typically found above the pectinate line, yet most commonly is a branch of the internal pudendal artery. It anastamoses with the superior rectal artery, inferior rectal artery, and the inferior vesical artery, and gives off an important branch to supply the prostate in some individuals.

The ventral urogenital sinus will form the bladder (except the trigone) and sex-specific structures. In the female, it also forms the urethra and vagina. In the male, it will form the prostate gland, prostatic urethra, and membranous urethra. The trigone is mesodermal in origin and forms from the Wolffian ducts.

CLINICAL SIGNIFICANCE

▍Haemorrhoids

These are vascular cushions that lie in the anorectal canal and have an important role in continence by protecting the internal and external anal sphincters. They are anastamoses of the vascular sinusoids between the superior and inferior rectal systems and can become engorged with blood, swollen, and inflamed in *haemorrhoid disease*.

Haemorrhoids are found at the left lateral, right anterior, and right posterior (3, 7, and 11 o'clock) positions. They can become pathological when other conditions cause increased intra-abdominal pressure and straining; these include constipation or diarrhoea, low-fibre diets, ascites, chronic coughs, obesity, prolonged sitting, and intra-abdominal malignancies. Haemorrhoids can be internal or external dependent on their position relative to the pectinate line.

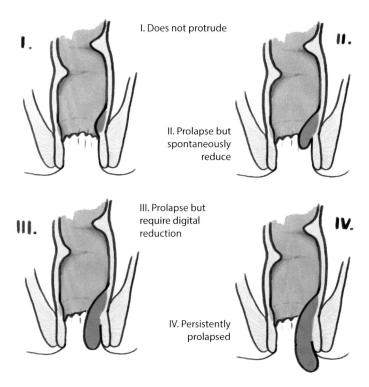

Figure 16.2 Grades of internal haemorrhoids.

External haemorrhoids are not graded, whereas internal haemorrhoids are classified by their location. This is because they can exist above the pectinate line but, when severe, prolapse to lie at or below the line (Figure 16.2). If minor, the patient presents with painless, bright red bleeding per rectum, that might be seen on the toilet bowl, surface of the stool, or when wiping. Note that the blood would be on the surface and not mixed in, which would be indicative of a malignancy or a gut inflammatory process. Haemorrhoids are associated with other symptoms such as pruritus, discharge, or sensation of fullness. The symptom of pain is associated with thrombosis of the haemorrhoid and requires analgesia or surgical excision.

The haemorrhoidal disease is managed according to its severity and recurrence. Acutely, if presenting within 72 hours, external haemorrhoids can be treated with topical analgesia or excised if they are particularly painful. For more chronic problems, haemorrhoid disease can be managed with non-operative changes such as weight loss, high-fibre diets or lifestyle modification (e.g. increased exercise). Where internal haemorrhoid disease is first- or second-degree with recurrent inflammation, then rubber-band ligation can be performed under direct vision with a proctoscope. More severe disease may require haemorrhoidal artery ligation or haemorrhoidectomy.

▌ Suprapubic Catheter

The bladder is a retroperitoneal structure and so can be accessed anteriorly for insertion of a suprapubic catheter. This may be used when access in not possible via the urethra, or when the bladder is reconstructed (e.g. with an ileal pouch).

▌ Hirschsprung's Disease

If there is a failure of neural crest cells to migrate or differentiate into the neurons of the enteric nervous system then this is known as *Hirchsprung's disease*. The lack of innervation leads to aperistalsis of the bowel followed by hypertrophy as the bowel is persistently contracted with no signal to relax. Typically, it is not the whole bowel which is affected, with *aganglionic segments* most commonly occurring in the sigmoid colon. The infant presents with either a distended abdomen due to the presence of a megacolon, or with failure to pass the first faecal movement (meconium). Unforgettably, on insertion of a finger into the newborn's anus for a digital rectal examination (to rule out an imperforate anus), there is the expulsion of an explosive bolus of stool.

▌ Imperforate Anus

This occurs due to inadaquate perforation of the anal membrane. It can occur anywhere in the anorectal canal (high or low) and presents with a failure to pass stool. It is associated with other genetic conditions including VACTERL syndrome. It is treated with surgery to create a colostomy to allow the stool to bypass the anus.

RELEVANT MOLECULES

- None

KEY POINTS

- The hindgut will form the the final third of the transverse colon, descending colon, sigmoid colon, rectum, and upper anal canal.
- Neural crest cells migrate to the bowel to form the enteric nervous system.
- The cloaca is a pouch at the end of the hindgut that will be separated by the urorectal septum into a ventral urogenital sinus and dorsal anorectal canal.
- The anal canal is separated into an upper and lower segment by the dentate line.
- Structures superior to this line are hindgut-derived.
- Inferior to the dentate line, structures are ectodermal in origin.
- This distinction leads to a difference in neurovascular supplies.

17 MESODERM: SPLEEN AND URINARY SYSTEM

MESODERMAL TISSUES AND ORGANS

The mesoderm forms key organs that lie between the endoderm-derived gut tube and the surrounding cavity. This concept can be appreciated by looking at the map of the trilaminar disc (Figure 17.1). The intermediate mesoderm forms the kidney, lower urinary tract, and reproductive system.

The lateral plate mesoderm forms the spleen, vessels, lymphatics, smooth muscle, and heart. It is divided by the *coelom* into the somatic and splanchnic mesoderm. The coelom creates the space between major organs and cavities (e.g. the space within greater/lesser sacs). The splanchnic mesoderm forms the organs (*splanchnic*

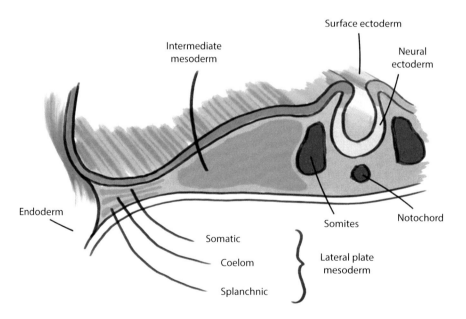

Figure 17.1 Germ layer map of the trilaminar disc.

144 EMBRYOLOGY EXPLAINED

means 'relating to the viscera/organs'), while the somatic mesoderm will form the peritoneal/pleural lining of the body's cavities (*somatic* means 'relating to the body').

DEVELOPMENT OF THE SPLEEN

Around the fifth week of development, the spleen begins to form from the splanchnic lateral plate mesoderm. These cells proliferate within the mesogastrium of the stomach (dorsal mesentery) to form the spleen. Due to its location within the mesogastrium, rotation of the stomach positions the spleen into the upper left quadrant of the abdomen. Due to this movement, two key structures form (Figure 17.2): first, the mesogastrium ventral to the spleen will become the

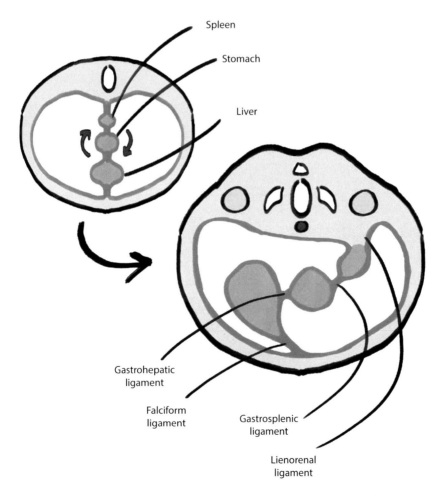

Figure 17.2 The rotation of the stomach leading to the final position of organs and their connecting ligaments.

gastrosplenic ligament; second, the mesogastrium dorsal to the spleen is brought into contact with the surrounding peritoneum and left kidney to form the *lienorenal ligament*. The spleen functions as a haematopoietic organ within the fetus until the third trimester, but forms only lymphocytes and monocytes later in life.

DEVELOPMENT OF THE URINARY SYSTEM

The intermediate mesoderm begins to organise and proliferate in the fourth week of development. It will form the kidneys, ureters, adrenal glands, gonads, and genital ducts. The development of the urinary system is a systematic process involving a staged progression through a series of structures: *pronephros, mesonephros,* and *metanephros*. These correspond to evolutionary stages of the kidney, with the pronephros forming the adult kidney in some primitive fish, the mesonephros in small mammals, and the metanephros in humans. It is important to understand that the urinary system will develop into two distinct segments: one that produces urine (excretory) and one that removes it (collecting).

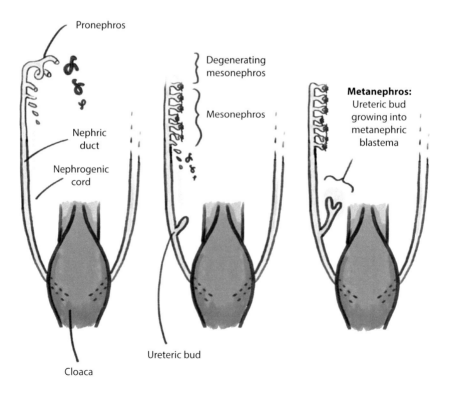

Figure 17.3 The developing kidney with formation of the pronephros, mesonephros, and metanephros.

Early in the fourth week of development, condensation of the intermediate mesoderm forms two *urogenital ridges* parallel to the midline. These ridges epithelialise at the cervical level to begin forming primitive rods – the *pronephric ducts*. Ventromedial to the rods, epithelial buds of *nephrotome* connect to the pronephric duct and communicate with vessels to create a very rudimentary kidney (Figure 17.3). This exists superiorly in the embryo and entirely regresses by the end of the fourth week.

The rods continue to grow caudally to form the *mesonephric (Wolffian) ducts*, which first appear in the thoracic and lumbar regions of the embryo. From the ducts, mesonephric tubules grow medially to meet small groups of capillaries; here, the tubule expands to surround the vessels with a *glomerulus* to form the earliest *Bowman's capsules*. This primitive system is capable of producing a diluted urine that is functional between weeks 6 and 10. The mesonephric ducts continue to grow caudally towards the cloaca, such that 40 tubules develop in total. However, most will regress and about 20 will remain in the male for development of the genitalia. Importantly, *the mesonephric (Wolffian) duct will completely regress in females.*

Around the fifth week of development, caudal to the mesonephros, the metanephros begins to develop into two separate components. A *ureteric bud* will branch from the metanephric duct. This will form the collecting segments of the urinary system including the ureters, renal pelvis, major and minor calyces, and collecting ducts/tubules. The bud grows into the *metanephric blastema*, which forms the excretory components of the urinary system and develops into the Bowman's capsule, proximal convuluted tubule, loop of Henle, and distal convuluted tubule. The metanephric blastema forms a cap around the growing ureteric bud. This bud elongates and bifurcates about 20 times in order to form the collecting ducts of 1–3 million nephrons (Figure 17.4).

Meanwhile, the metanephric cap will respond to signals from the distal convuluted tubules to form *renal vesicles* – round collections of cap tissue. These vesicles will elongate to form an S-shaped tubule that communicates proximally with the glomerular capillaries as a Bowman's capsule, and distally with the collecting duct as the *distal convuluted tubule*. This is now a complete nephron that drains into the minor calyces. This kidney is functional from the tenth week of gestation and generates urine to contribute to the amniotic fluid surrounding the fetus. Recall that the metanephros is the caudal end of the duct (towards the cloaca), so the kidneys later ascend into the lumbar region to join the adrenals. As they rise through the fetus, they acquire new blood supplies. The ureters will descend and gain vascular supply as they do so; they will attach to the posterior wall of the bladder (developed from the ventral urogenital sinus) and form the trigone.

The development of the kidney is dependent on upregulation of *WT1* in the metanephric blastema, and *fibroblast growth factor 2 (FGF2)* and *bone morphogenetic protein 7 (Bmp7)* signalling in the ureteric bud. Pax2 and Wnt proteins are expressed in the metanephric blastema to form the vesicles and tubules. There is constant

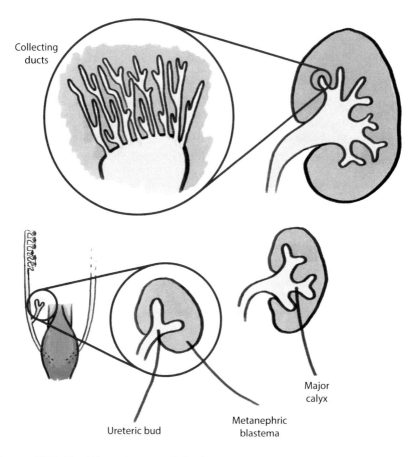

Figure 17.4 The bifurcating ureteric bud.

communcation between the bud and cap through these molecules to ensure that the correct structures both form and adhere to one another.

DEVELOPMENT OF THE ADRENAL GLANDS

Around the fourth week of development, intermediate mesoderm cells from the urogenital ridge will begin to proliferate. These cells will differentiate into a fetal adrenal cortex in the sixth week that will produce cortisol by the eighth week. This *cortex* continues to develop into birth and beyond, with the zona glomerulosa and fasciculata appearing late in pregnancy/soon after birth, and the reticularis developing in early childhood. The *medulla* develops following migration of neural crest cells in the seventh week. These cells lie at the medial border of the developing adrenal cortex and are only enveloped by the cortex in late pregnancy.

CLINICAL SIGNIFICANCE

■ Disorders of the Spleen

The pathologies of the spleen exist on a spectrum from absence to accessory tissue.

In *asplenia*, the spleen is completely absent. This is very rare but usually occurs as part of wider issues with mesoderm-derived tissues (e.g. cardiovascular disorders). Where asplenia happens in isolation, the child develops a primary immunodeficiency and there is a high risk of life-threatening bacterial meningitis or sepsis. Patients with *congenital hyposplenia* also face these risks due to underdevelopment of the splenic tissue. This population must be treated the same as post-splenectomy patients with a full course of vaccinations, particularly against encapsulated bacteria (*Pneumococcus*, *Meningococcus*, and *Haemophilus*).

The most common variance occurs in 10–20% of the population and involves additional nodules of splenic tissue, known as *accessory spleens*. They can be found anywhere within the abdomen but most commonly occur along the path of the splenic vessels (the surrounding omentum, mesentery, or ligaments), or towards the gonads. It is important to understand these phenomena as they can be identified during operations or on computed tomography imaging.

■ Horseshoe Kidney

Fusion of the inferior poles of the kidneys leads to the formation of a horseshoe kidney that lies anterior to the aorta. As it ascends, it becomes trapped by the inferior mesenteric artery and cannot move further (Figure 17.5).

■ Supernumerary Arteries

Each kidney can have up to three arteries supplying it. They will have the renal artery branch of the aorta as their main supply, but may bring an *accessory vessel* from migration. This occurs in about 25% of patients and is important for identification in transplant operations. If the accessory artery perforates the substance of the kidney, rather than entering at the hilum, then it is known as an *aberrant renal artery*. This is more common in horseshoe kidneys. Typically, aberrant arteries branch off the aorta to supply inferior poles.

■ Duplicated Kidney and Urinary Tract

This condition occurs when the ureteric bud prematurely bifurcates prior to penetrating the metanephric blastema to form a cap. It can lead to the formation of an accessory kidney, or additional urinary tracts. Neither of these confer any renal advantage but, rather, increase the susceptibility of the individual to urinary tract infections.

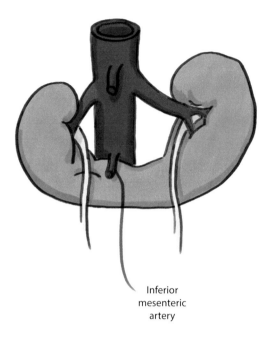

Figure 17.5 A horseshoe kidney unable to ascend beyond the inferior mesenteric artery.

▌ Pelvic Kidney

A pelvic kidney occurs when there is a failure of the kidney to ascend and it remains in the pelvis. This is largely asymptomatic but can create greater difficulty in diagnosing renal pathology as the symptoms present atypically.

▌ Renal Agenesis and Potter's Sequence

In this condition, there is complete failure to form renal tissue. It can be unilateral or bilateral, and occurs due to a lack of interaction between the ureteric buds and the metanephric blastema. This leads to *oligohydramnios* due to a reduction in the production of amniotic fluid. As a result of the reduced fluid, there is less space between the fetus and the mother's abdominal cavity. This leads to increased pressure on the embryo's body, which causes characteristic features of compression (also known as *Potter's sequence*) such as a sloped forehead, flattened ('parrot beak') nose, shortened fingers, and compression of internal organs leading to hypoplasia. It also leads to poor development of the lungs as the amniotic fluid is required for lung development.

▌ Wilms' Tumour

Also known as a *nephroblastoma*, this condition occurs due to a mutation in the WT1 (or WT2) gene. It is characterised by consisting of three types of cellular tissue: metanephric blastema, mesenchyme/stroma, and epithelium. It is diagnosed by the presence of a large, often painless, abdominal mass in the infant.

▌ Renal-Coloboma Syndrome

Mutations in the PAX2 gene lead to defects in the kidneys and the eye. The mesenchymal blastema requires this gene to form the correct structures, and its modification leads to renal hypoplasia and vesicoureteral reflux. The latter occurs when there is an inappropriate connection between the ureter and the posterior bladder wall. Finally, a coloboma is a defect in the eye (iris, retina, and/or optic nerve), most commonly a slit in the iris leading to disfigurement of the shape of the pupil. This occurs because PAX2 is important in the fusion of ventral parts of the eye.

▌ Polycystic Kidney Disease

In this condition, there is dilatation of the excretory component of the urinary system – particularly in the tubules – leading to multiple cysts on the kidney. The most common cause of this is an autosomal dominant mutation in the *PKD1* gene. Less frequently, it can be due to an autosomal recessive mutation in the *PKD2* gene. These two genes produce cilia on renal cells, but the mechanism for how this leads to cysts is less clear. It is thought to involve calcium regulation.

RELEVANT MOLECULES

- *WT1:* when upregulated in the metanephric blastema, this gene makes it responsive to the ureteric bud
- *FGF2* and *BMP7:* genes expressed in the ureteric bud whose signalling proteins act in metanephric blastema development
- *PAX2* and *WNT:* genes expressed in the metanephric blastema to induce formation of the tubules

KEY POINTS

- The intermediate mesoderm will form the kidney, lower urinary tract, and reproductive system.
- The lateral plate mesoderm forms the spleen, vessels, lymphatics, smooth muscle, and heart.
- Development of the urinary system is a defined process involving staged progression through a series of structures: pronephros, mesonephros, and metanephros.
- The pronephros regresses entirely in early development.
- The mesonephric (Wolffian) duct will completely regress in females.
- The metanephros is the definitive kidney structure in the embryo.
- Urine produced by the kidney contributes to the amniotic fluid and is important for embryogenesis.
- Development of the kidney is dependent on WT1 gene upregulation in the metanephric blastema.

18 MESODERM: INTERNAL AND EXTERNAL GENITALIA

It is important to know the individual components of sex determination within the embryo in order to understand how the processes occur. The key concepts to comprehend are:

- *Genetic sex:* the chromosome configuration (karyotype) of the embryo (e.g. XX or XY)
- *Gonadal sex:* the characterisation of the gonads that form within the embryo (e.g. testes)
- *Primordial germ cells* (*PGCs*): these are the undifferentiated stem cells that will become either spermatozoa or oocytes
- *External genitalia:* the sex organ of the embryo (e.g. penis)
- *Internal gonads:* an organ that produces gametes (e.g. ovary)
- *Gender:* not the same as genetic or gonadal sex; this refers to the individual's identification of self (e.g. man, woman, non-binary) and is not determined in the embryonic stage of development

The gonads form from the intermediate mesoderm of the paired urogenital ridges. The genitalia develop from the mesonephric (Wolffian) ducts in males, and from the *paramesonephric* (*Müllerian*) *ducts* in females (Figure 18.1). The gonads and genitalia are indistinguishable in males and females until the seventh week of development when the genetic sex influences their development; as such, the presumptive gonad tissue is considered *bipotential*, in that it can form either male or female gametes.

In XY individuals, the *sex-determining region Y* (*SRY*) *gene* leads to the formation of male gonads/genitalia, with its absence leading to the development of female gonads/genitalia. The PGCs appear in the fourth week of development and are derived from the epiblast. They begin to migrate during hindgut development and travel through the dorsal mesentery towards the urogenital ridges to lie medial to the developing ducts.

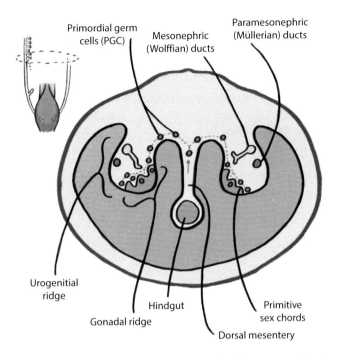

Figure 18.1 The primordial gonadal structures within the urogenital ridge.

The gonads will develop from the intermediate mesoderm (mesenchyme), overlying coelomic epithelium (mesoderm/mesothelium on the posterior abdominal wall), and the PGCs. The mesoderm and mesothelium proliferate to form the urogenital ridges, with a segment of these forming distinct *gonadal ridges* in the fifth week. Finger-like extensions of the coelomic cells, known as *primitive sex cords*, grow inwards from the gonadal ridges. These ridges now have a surrounding cortex and an inner medulla. The testis will develop from the medulla, whereas ovaries are created from the *cortex*. As these cords form, the paramesonephric ducts emerge laterally.

DEVELOPMENT OF THE MALE GONADS

The SRY gene upregulates the autosomal gene for transcription factor SOX9, which in turn promotes the differentiation of the male gonadal cells by upregulating the production of *steroidogenic factor 1 (Sf1)*. SOX9 generates a positive feedback cycle for itself by stimulating *fibroblast growth factor 9 (FGF9)*, which upregulates Sox9 and causes the mesonephric tubules to grow towards the *rete testis*.

In the sixth and seventh weeks of development, the XY-containing PGCs invade the *primary sex cords* and release *testis-determining factor*. The primary sex cords will grow into the medulla and differentiate into the *testis cords* (seminiferous tubules) and *Sertoli cells*. The parts of the cord deepest within the medulla (deficient in PGCs) will develop into the rete testis cords, while those in the most peripheral medulla form the *tunica albuginea* (Figure 18.2). The cortex will then degenerate. The rete testis will open into the most caudal parts of the mesonephric duct. The *testis cords* actually remain a rod until puberty, where the increase in testosterone levels leads to its canalisation and the formation of seminiferous tubules, capable of transporting sperm from the testis to the ejaculatory ducts.

The Sertoli cells respond to SF1 signals to secrete *anti-Müllerian hormone* (*AMH*, also known as Müllerian-inhibiting substance), which has a triple function: the first is to suspend the germ cells as spermatogonia in *meiotic arrest* until puberty; the second is to stimulate intermediate mesodermal cells to differentiate into *Leydig cells*; and the final and most important function is to suppress the paramesonephric ducts so that they regress.

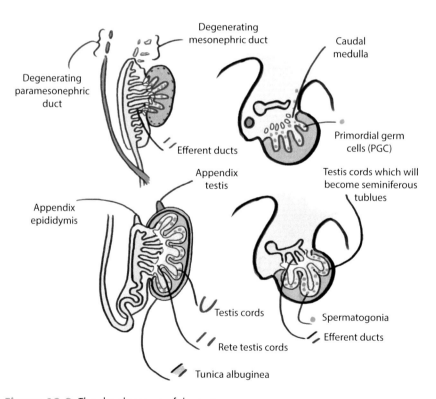

Figure 18.2 The development of the testes.

DEVELOPMENT OF THE MALE GENITAL DUCTS

The male genital ducts form in two parts: cranially from the mesonephric ducts and caudally from the *urogenital sinus*. The Leydig cells secrete testosterone, which acts locally to influence genital development. It promotes the persistence of the mesonephric ducts and induces the connection between the rete testis and caudal-most 5–12 mesonephric tubules (bringing together the two systems). These drain into the mesonephric duct, which later becomes the *epididymis*, *seminal vesicles*, and *vas (ductus) deferens*. The ejaculatory ducts form where the vas deferens opens into the posterior wall of the urogenital sinus, at a site known as the *verumontanum (seminal colliculus)*.

There are two vestigial remnants of the regressed paramesonephric ducts in males: the *appendix testis* (a small cap of tissue on the superior pole of the testis) and the *prostatic utricle* (an expansion on the prostatic urethra, used as a landmark in rigid cystoscopy). The utricle is likely homologous to the vagina, while the seminal colliculus is supposedly homologous to the hymen.

The prostate forms in the tenth week of gestation as an endodermal outgrowth of the pelvic urethra (from the posterior urogenital sinus). Its development is stimulated by *dihydrotestosterone (DHT)*, which is generated from testosterone after conversion by *5-α-reductase*. The bulbourethral glands will form in a similar manner, by emerging as paired endodermal outgrowths from the membranous urethra (which is formed from the urogenital sinus).

DESCENT OF THE TESTES

The testes develop in the lumbar region of the abdomen and descend to the groin area by utilising the contracture of tissues. At around the second month of development, the testis is attached to the posterior abdominal wall by a urogenital mesentery; this becomes ligamentous, transforming into the *caudal genital ligament* with a mesenchymal component known as the *gubernaculum*. It connects the caudal pole of the testis to the labioscrotal swelling developing outside of the abdominal cavity. Ahead of the testis is a fold of peritoneum known as the *vaginal process* that follows a similar path through the inguinal region to the labioscrotal swelling. These structures are important as the abdominal wall muscles grow around them to form the inguinal canal.

The descent of the testes is a combination of two factors: first, there are contractile proteins within the gubernaculum that actively move the testes; and secondly, there is a greater rate of growth of the abdominal cavity than of the gubernaculum, such that the testis (anchored by the ligament to the labioscrotal swelling) moves caudally. The active gubernaculum component of this descent is responsive to testosterone and DHT signalling.

DEVELOPMENT OF THE FEMALE GONADS

If the SRY gene is not present, then female gonads will form as the default for the embryo. In the absence of the SRY gene, the primary sex cords lose their structure and form loose clusters of mesenchymal cells within the medulla; these will later disappear to be replaced by highly vascular stroma (Figure 18.3). Meanwhile, the PGCs continue to proliferate within an expanding cortex into oogonia. Many of these oogonia apoptose and become reabsorbed into the cortical epithelium. This expands the size of the cortex to give rise to *cortical cords* that split into small clusters surrounding each individual *oogonium* as a *follicular cell*. By the twentieth week of development, the oogonia have entered a suspended prophase I of meiosis to become primary oocytes.

This developmental process is dependent on activation of the *DAX1* gene. Owing to the lack of Sox9 production (due to no *SRY* gene), *Wnt4* is not inhibited and its upregulation promotes the expression of DAX1. This in turn inhibits any expression of SOX9, in order to safeguard formation of the ovaries.

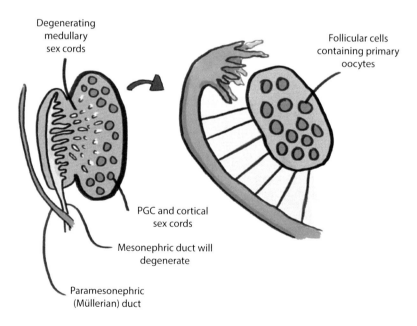

Figure 18.3 The development of the ovaries.

DEVELOPMENT OF THE FEMALE GENITAL DUCTS

Once the ovaries have formed, the mesonephric ducts will regress to remain only as vestigial *Gartner's ducts*. This leaves the paramesonephric ducts and urogenital sinus to form the uterus and vagina.

In the eighth week of development, the two paramesonephric ducts will fuse in the midline (Figure 18.4), bringing with them the peritoneal folds that form the *broad*

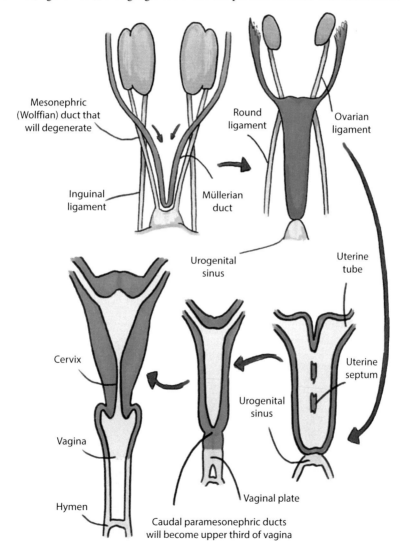

Figure 18.4 The formation of the uterus and vagina.

Mesoderm: Internal and external genitalia

ligaments. Surrounding mesenchymal cells invade the primordial uterus to form the *myometrium* and *endometrium*. This process generates the presumptive uterus suspended within the ligaments that then grows towards the urogenital sinus. As the paramesonephric ducts make contact with the posterior urogenital sinus, two *sinovaginal bulbs* emerge from the sinus to form a solid *vaginal plate*.

Starting in the eleventh week and completing in the fifth month of development, this plate will canalise to complete the uterovaginal canal. The upper third of the vagina (the vaginal fornices) will be formed from the paramesonephric ducts, while the lower two-thirds develops from the urogenital sinus (Figure 18.4). Furthermore, *vestibular (Bartholin's) glands* develop as outgrowths from the urogenital sinus. Meanwhile, the *Fallopian tubes* develop as unmerged cranial aspects of the paramesonephric ducts with the cranial opening forming the *fimbrae*. Following regression of the mesonephric ducts and growth of nearby structures, the Fallopian tubes become horizontal and the ligament connecting the mesonephric ducts and the paramesonephric duct becomes the *suspensory ligament of the ovary*. The opening of the vagina is marked by a *hymen*, a thin piece of tissue. The much shorter female urethra is generated by ventral components of the urogenital sinus; its shorter length makes females more susceptible to urinary tract infections from the flora in the surrounding perineal areas.

DEVELOPMENT OF THE INDIFFERENT EXTERNAL GENITALIA

The external genitalia initially develop in a uniform fashion from all three germ layers in both sexes: the ectoderm forming overlying skin; the lateral plate mesoderm forming the genital swellings; and the endoderm for urethral components. The swellings that surround the cloacal and urogenital membrane are indistinguishable until about the tenth week of development, thus it is known as the *indifferent stage*.

Cloacal folds will suround the cloacal membrane to meet cranially at a *genital tubercle* (Figure 18.5), which will lengthen early to form a phallus-shaped structure that pulls caudal structures upwards. When the urorectal septum meets the cloaca and divides the cloacal membrane, the cloacal folds become *urethral folds*. Lateral to these, *genital swellings* appear to form the *labioscrotal swellings*. The urogenital membrane will narrow to form the *urethral meatus*. At this stage, the male and female external genitalia are indistinguishable, but these three core structures will form different structures in each sex, as described in Table 18.1.

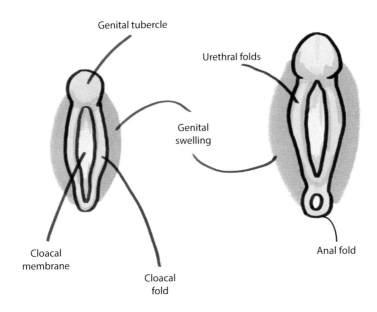

Figure 18.5 The development of external genitalia in the indifferent stage.

Table 18.1 Developmental fate of the genital tubercle, genital folds, and genital swellings in males and females

Sex	Genital tubercle	Genital folds	Genital swellings
Male	Body and glans of penis Corpora cavernosa Corpus spongiosum	Ventral penis Penile raphe	Scrotum Scrotal raphe
Female	Clitoral body and glans	Labia minora	Labia minora Mons pubis

DEVELOPMENT OF THE MALE EXTERNAL GENITALIA

All three germ layers are involved in the formation of the external genitalia. Under the influence of DHT, the external genitalia begin to undergo *masculinisation* (Figure 18.6). The genital tubercle (the *phallus*) elongates further, pulling the urogenital membrane (the presumptive urethral meatus) cranially to form the

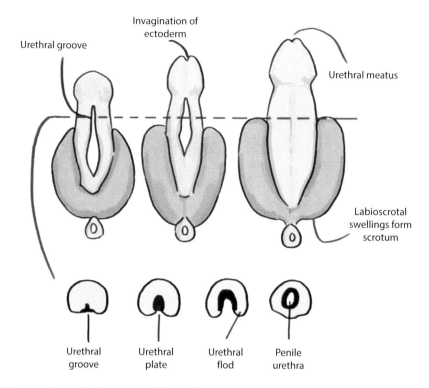

Figure 18.6 The formation of the male external genitalia.

urethral groove. The phallus expands cranially to form the *glans*. As such, there is now a urethral groove caudal to the phallus. This groove becomes lined with endoderm to form the *urethral plate*.

This endodermal plate brings new cells in to differentiate the urethra from surrounding genital tissue. Its introduction allows the urethral folds to close over the urethral groove in order to form the *penile urethra*. At the tip of the glans, an invaginating pit of ectoderm generates the external urethral meatus and the distal-most segment of urethra. Finally, the labioscrotal swellings will meet in the midline to form the scrotum.

DEVELOPMENT OF THE FEMALE EXTERNAL GENITALIA

In females, the external genitalia begin to develop in response to secreted oestrogens (Figure 18.7). The genital tubercle elongates only slightly to become a phallus, which then forms the *clitoris* with erectile tissue (analogous to the cavernosum). The urogenital groove that lies between the urethral folds will remain open to contain the *urethral opening* ventrally/cranially and the *vaginal opening* dorsally/caudally, with the remainder forming the *vestibule* between the two. Meanwhile, the urethral folds do not overgrow the groove, but rather form the *labia minora* lateral to the vaginal opening. The labioscrotal (genital) swellings will form the *labia majora*.

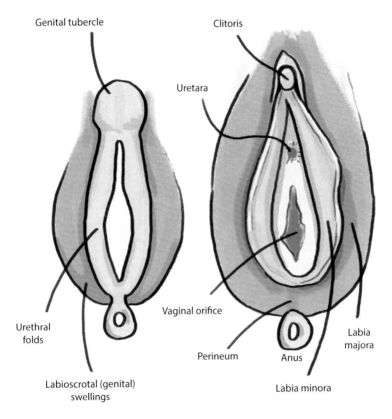

Figure 18.7 The development of the female external genitalia.

CLINICAL SIGNIFICANCE

When considering the clinical syndromes associated with the formation of gonads and genitalia, it is useful to recall the initial descriptors of the different forms of sex: genetic, gonadal, internal genitalia (genital ducts), and external genitalia. This will help learners to understand the genetic and molecular pathways involved in each of the configurations. The terminology surrounding these conditions has changed to better reflect the underlying pathophysiology. Nomeclature such as '*intersex*' and '*hermaphroditism*' have been replaced with '*differences of sex differentiation*' (DSD), which assess the development from an underlying genetic perspective as '46XY', '46XX' and 'sex chromosome' DSDs. Genitalia that lie between that of typical female or male genitalia are often described as '*ambiguous*'.

▪ Turner Syndrome (Sex Chromosome DSD)

Most individuals carry 46 chromosomes of which 22 pairs are the autosomal chromosomes and one pair are the sex chromosomes. Genetic males have the configuration 46XY, and genetic females the configuration 46XX. In *Turner syndrome*, individuals have a genetic condition where the second X chromosome is either partly or completely missing, denoted as '45X0' (note that 45Y0 would not be compatible with life).

The classic presentation is females with a webbed neck, low-set ears, and short stature. The effect on fertility is dependent on the specific genetic mutation and proportion of the second X-chromosome deletion. Almost all individuals are infertile due to ovarian dysgenesis, with only a small number of PGCs developing into oocytes. This also means that most of these individuals have *primary amenorrhoea* (failure to establish menstruation) due to a lack of hormones produces by ovarian tissue. However, through hormone therapy (with oestrogen) and *in vitro* fertilisation, individuals are able to undergo puberty and carry a pregnancy. This is because the primary reproductive defect is with the sex chromosome and PGCs that influence the development of the gonads, and not with the internal female reproductive system (e.g. uterus).

▪ Klinefelter Syndrome (Sex Chromosome DSD)

Individuals with this condition have an additional X chromosome (from the father) so carry the 47XXY karyotype. Individuals are male and generally asymptomatic, but will have infertility and hypoplasia of the testicles. At puberty, they may notice gynaecomastia, sexual anhedonia, or less hair growth. Generally, however, it is associated with a higher incidence of many other inflammatory and autoimmune conditions. The hypogonadism occurs due to the lower levels of testosterone caused by the higher levels of follicle-stimulating hormone (FSH) and luteinising hormone (LH) associated with the extra X chromosome. The exact mechanism is not well understood, and it is not as simple as a hormone imbalance as it has been identified that the additional chromosome affects autosomal gene expression also. However, modern artificial insemnation and sperm extraction techniques have meant that these patients can conceive children.

▌ Androgen Insensitivity Syndrome (46XY DSD)

Also known as *'testicular feminisation syndrome'*, androgen insensitivity syndrome (AIS) occurs in individuals whose Leydig cells produce testosterone but without the ability of androgen receptors to bind the testosterone/DHT. This leads to the formation of undescended testes and female external genitalia. This is because the testes form in response to SRY gene expression, while the external genitalia are unresponsive to the DHT and default to female development pathways. The lack of masculinisation of the external genitalia is described as *feminisation*. The insensitivity continues to puberty as there is a lack of response to testosterone.

It is important to note, however, that this condition exists on a spectrum relative to the extent of unresponsiveness to testosterone: complete, partial, and mild. In *complete AIS*, there is total feminisation of the external genitalia; in *partial AIS* there is incomplete masculinisation; and in *mild AIS* the genitalia will appear male. In all cases, the Sertoli cells continue to produce AMH, which will lead to the regression of the paramesonephric ducts; therefore, there is no formation of oviducts, uterus, or the upper third of the vagina. Without the testosterone response, the male genital duct derivatives (e.g. the epididymis, vas deferens, and seminal vesicles) are typically absent.

▌ 5-α-Reductase Deficiency (46XY DSD)

This condition leads to a very similar clinical presentation to AIS, with internal male gonads and external ambiguous or female genitalia. This enzyme is responsible for the conversion of testosterone to DHT, which is required for signalling.

Interestingly, however, the mesonephric ducts and their derivatives persist as they require testosterone (not DHT) to continue development. The external genitalia may have the appearance of a *micropenis* with *hypospadias* (due to an enlarged clitoris with an inferior-lying urethral opening). These individuals may have feminine features in childhood but then undergo a male puberty because the testes release testosterone – with the resulting descent of the testes, deepening of the voice, and male-pattern hair development. As such, in some cultures where the incidence of this deficiency is more common, the entire process is celebrated with a marked occasion.

Furthermore, the release of testosterone can lead to lengthening of the clitoris/micropenis to form a phallus that may appear superficially like a penis. In these individuals, a hypospadias repair can be completed in order to place the external urethral opening on the tip of the phallus.

▰ Cryptorchidism (Undescended Testes)

Up to 4% of full-term infant males will be born with unilateral or bilateral undescended testes. It is not treated immediately as the descent of the testes can continue into infancy. On examination, the healthcare practitioner needs to determine if the testes are *impalpable* (not found), *palpable* (perhaps in the inguinal or intra-abdominal regions), or *retractile* (appearing in the scrotum temporarily). It is important to monitor and appropriately manage these infants as untreated cryptorchidism leads to decreased fertility and increased risk of testicular malignancy (due to the inappropriate physiological temperature and environment).

If the testes have not descended by the age of 6 months, then a referral for surgical management (*orchidopexy*) should be completed. Post-surgically, individuals with unilateral cryptorchidism have near-normal rates of fertility, whereas in those with bilateral cryptorchidism the fertility rate is reduced to 50%.

▰ Persistent Müllerian Duct Syndrome (46XY DSD)

In persistent Müllerian duct syndrome, there is either a mutation in the gene coding for production of AMH or in the gene encoding its receptor. The individual will develop male gonads and male external genitalia; however, there will be no regression of the paramesonephric ducts, leading to the formation of a small uterus, Fallopian tubes, and upper third of the vagina. Due to the development of the uterus, a broad ligament forms which impedes the descent of the testes, leading to cryptorchidism. An operation can be performed to remove the Müllerian structures and correctly position the testicles.

▰ Processus Vaginalis, Inguinal Hernias, and Hydroceles

As discussed earlier, the processus vaginalis is the fold of peritoneum that guides the inguinoscrotal descent of the testes. It will normally become obliterated after birth, but persistence can lead to the formation of an *inguinal hernia* or *hydrocele*.

With inguinal hernias, a sufficiently wide patent processus vaginalis provides a route through the inguinal canal for an indirect inguinal hernia, such that bowel, mesentery, or intra-abdominal fat can pass through the superficial and deep rings to the scrotum. It presents with a swelling inferior and lateral to the *pubic tubercle* of the pubic bone (since the inguinal ligament attaches to the pubic tubercle medially with an opening lateral to this attachment site); this distinguishes it from a *femoral hernia* which lies superior and medial to the pubic tubercle. Inguinal hernias are at risk of strangulation following incarceration, so are usually operated on to avoid acute presentation of ischaemia/necrosis.

Hydroceles are benign swellings of the scrotum caused by the accumulation of serous fluid within the patent processus vaginalis. They can be visualised on ultrasound, and on examination will characteristically *transilluminate* when assessed with a torch. They can occur in up to 3% of births but are asymptomatic, and only require treatment due to a secondary pathology such as infection.

▌ Hypospadias

This occurs due to the incomplete fusion of the urethral folds to enclose the urethral plate. It leads to a urethral orifice anywhere on the ventral/inferior aspect of the shaft of the penis. It can be corrected with surgery.

▌ Congenital Adrenal Hyperplasia (46XX DSD)

Congenital adrenal hyperplasia occurs as a result of defects in the enzymes for the formation of *glucocorticoids* and *mineralocorticoids*, most commonly *21-hydroxylase deficiency*. This means that all of the *17-hydroxyprogesterone precursor* molecule is shunted to production of sex hormones (e.g. testosterone), rather than other steroids such as cortisol. The absence of other forms of steroid hormones means that there is limited negative feedback, leading to high levels of adrenocorticotropic hormone that stimulate the *adrenals* and cause *hyperplasia*.

Since there is no SRY gene, the individual will form female gonads (ovaries). The effect on the external genitalia is dependent on whether *androstenedione* (formed directly after the 17-hydroxyprogesterone on the sex hormone pathway) will be converted into testosterone or oestrogen compounds.

However, androstenedione itself is a potent virilising agent, and if there is overwhelming production of androstenedione then this will lead to virilisation (masculinisation) of the external genitalia. These individuals will have no male genital ducts as the mesonephric ducts regress due to a relative lack of testosterone; the lack of the SRY gene also means that no Sertoli cells (or AMH) are produced and so the female genital ducts (e.g. the uterus) remain *in situ*.

On birth, these genitalia will appear as a scrotum with no palpable testes, so must be distinguished from cryptorchidism. Individuals with this condition will undergo female puberty, with the production of oestrogen. As such, reassignment surgery can be undergone to feminise the genitalia, if desired.

▌ Ovotesticular Disorder (46XX/XY DSD)

Also known as *'true hermaphroditism'*, this condition is very rare and involves the presence of both male and female gonads with ambiguous external genitalia. Using traditional terminology, all previously described conditions in this chapter with incongruency between internal and external genitalia would be considered *'pseudohermaphroditism'*.

Most commonly, *ovotesticular disorder* is due to a division of the ovum into two separate haploid ova, which are then fertilised by different XX/XY sperm cells; these two fertilised ova (zygotes) then fuse together early in development. Alternatively, it may be due to a genetic mutation where there is mosaicism of the Y chromosome onto the X chromosome (translocation of the SRY gene).

▌ Uterine Disorders

Malformations of the uterus can occur due to issues with the fusion of the paired paramesonephric ducts, and leads to many anatomical variants.

In *uterus didelphys*, there is duplication of the uterus due to a lack of fusion between the two ducts (Figure 18.8). These individuals can have two cervices and two vaginas, but will only have one horn from each uterus that links to the ipsilateral Fallopian tube and ovary. Some studies have found this condition to affect fertility. However, it is possible for each uterus to carry an individual pregnancy, with historical cases of

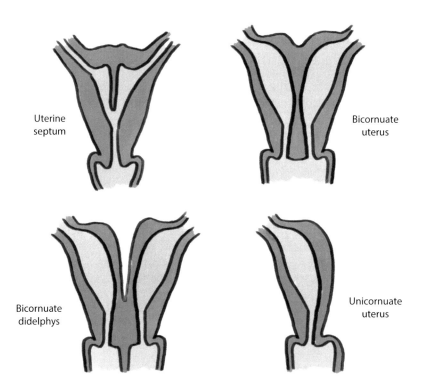

Figure 18.8 Uterine disorders.

triplets in the UK (twins in one uterus and single pregnancy in the other). Pregnant individuals with this condition require greater monitoring and most often require a Caesarean section. Like most disorders of the uterus, it is associated with higher complications of pregnancy including preterm birth, malpresentation, placenta praevia (placenta attached near or over the cervical opening), retained placental products, and premature rupture of membranes.

A *bicornuate uterus* forms when the upper portions of the paramesonephric ducts do not fuse, while the lower portions do. These lead to a uterus with a septum extending from the uterine fundus to the cervical os, the extent of which is used for its classification. As before, it requires an increase in surveillance of pregnancy with a higher risk of complications.

A *unicornuate uterus* occurs due to one of the paramesonephric ducts regressing or not forming appropriately. This uterus will communicate with the unilateral Fallopian tube and ovary. It is managed with the same risks as the other outlined uterine disorders.

RELEVANT MOLECULES

- *SRY:* sex-determining region of the Y chromosome that leads to the masculinisation of gonads, genital ducts, and genitalia in XY individuals
- *SF1:* a transcription factor that is required in the development of male gonads; it stimulates Sertoli cells to produce AMH
- *SOX9:* a gene and transcription factor upregulated by SRY that leads to the formation of the testes
- *FGF9:* promoted by SOX9, this gene encourages a positive feedback cycle to self-propagate the production and expression of SOX9
- *AMH:* produced by the Sertoli cells of the male genital ducts, this agent inhibits the paramesonephric (Müllerian) ducts in males, leading to their regression
- *Testosterone:* secreted by the Leydig cells, this hormone has local effects on the maintenance of the mesonephric (Wolffian) ducts and distant effects on the virilisation of external genitalia
- *DHT:* formed from testosterone, this stimulates tissues locally to form the male genital ducts and the external genitalia
- *5-α-reductase:* the enzyme required to convert testosterone into DHT

KEY POINTS

- Genetic sex refers to the embryo's karyotype, which can be different from its gonadal sex.
- Gonads develop from the intermediate mesoderm of the urogenital ridges.
- Gonads are indistinguishable until about the seventh week of development.
- Genitalia are indistinguishable until about the tenth week of development.
- The SRY gene on the Y chromosome in genetic males will differentiate the bipotential genital primordia into male genitalia.
- Spermatozoa and oocytes form from the PGCs.
- The mesonephric ducts persist in males.
- The paramesonephric ducts persist in females.
- Sertoli cells secrete AMH.
- Leydig cells secrete testosterone.
- The testes initially form in the lumbar region and undergo descent into the scrotum.